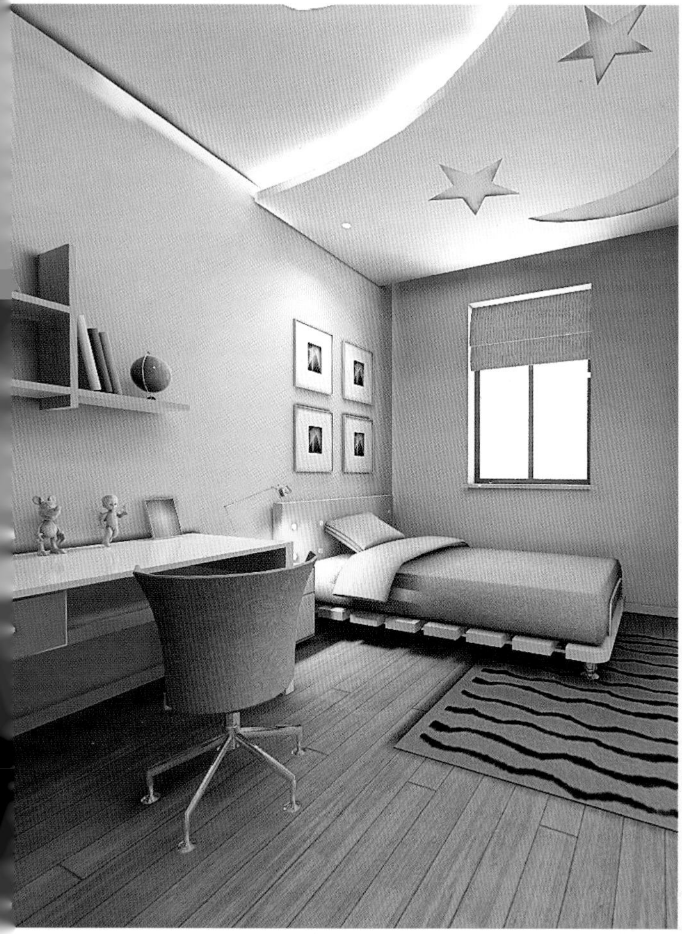

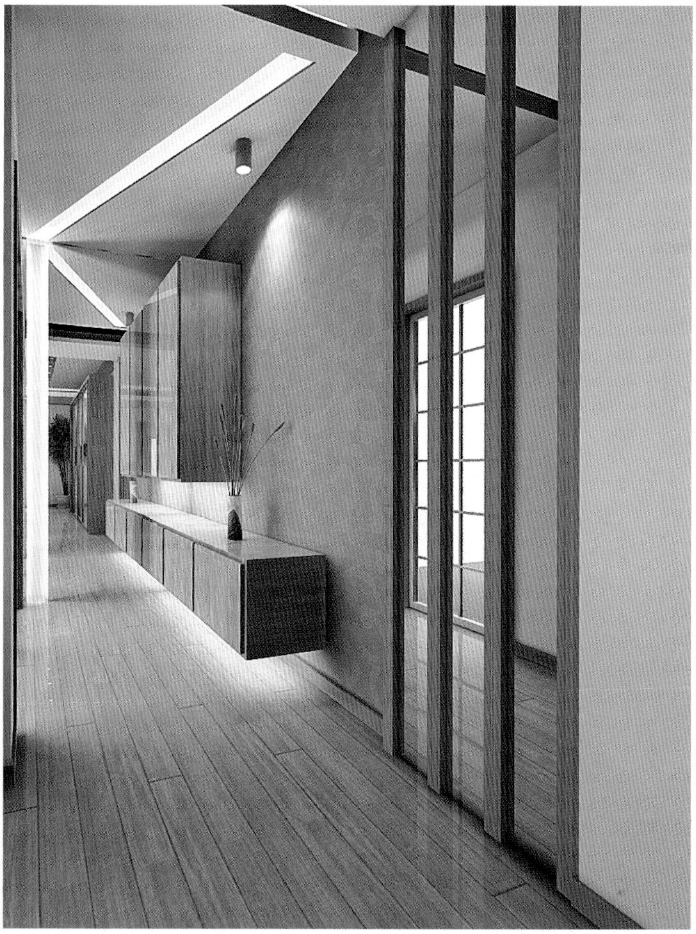

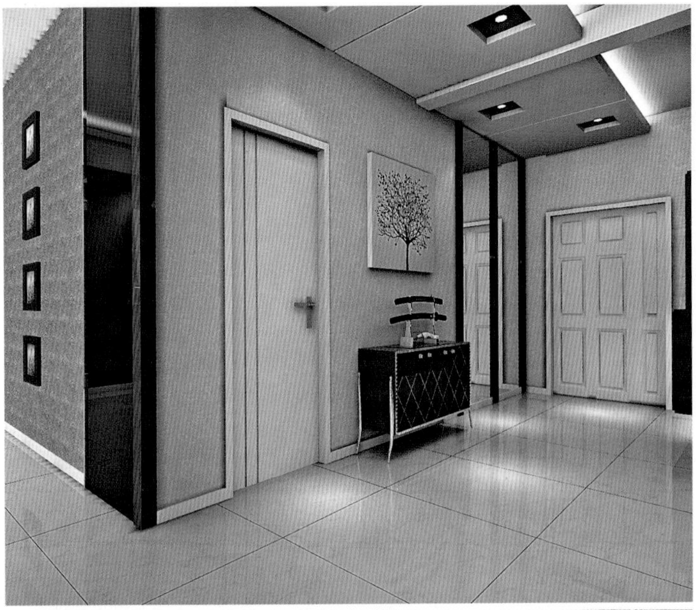

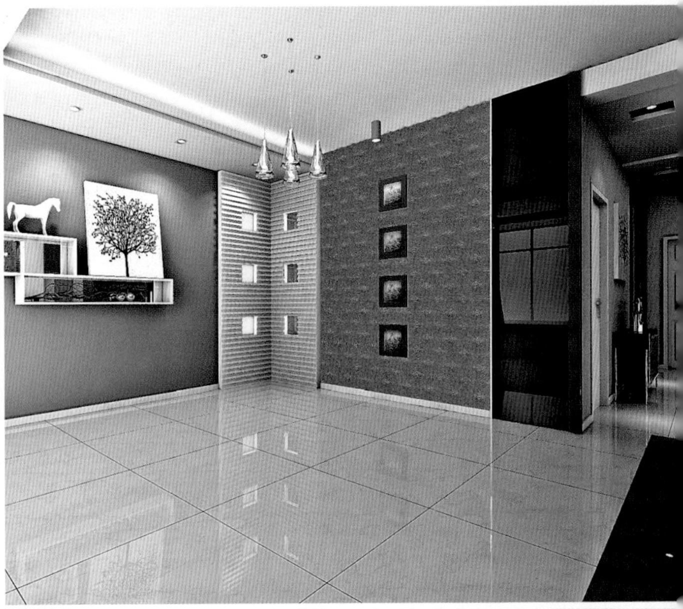

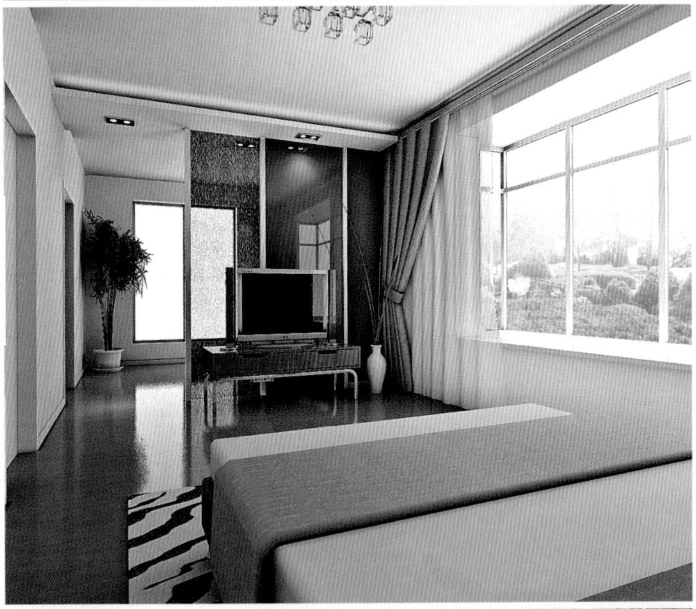

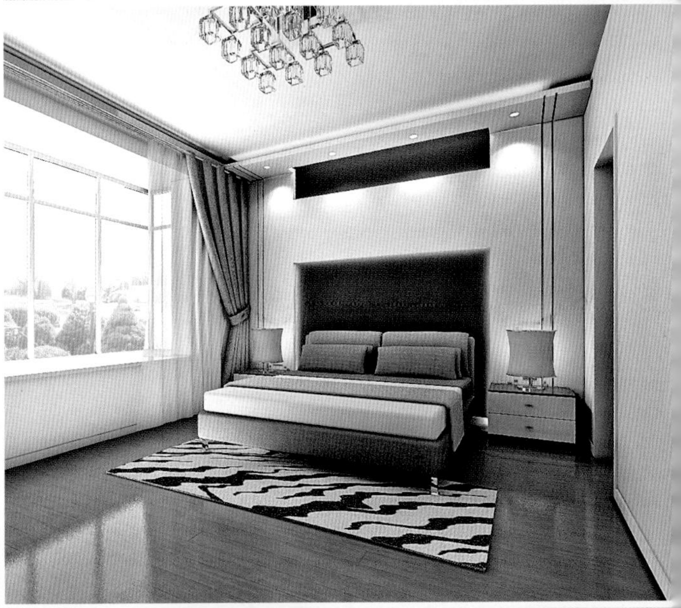

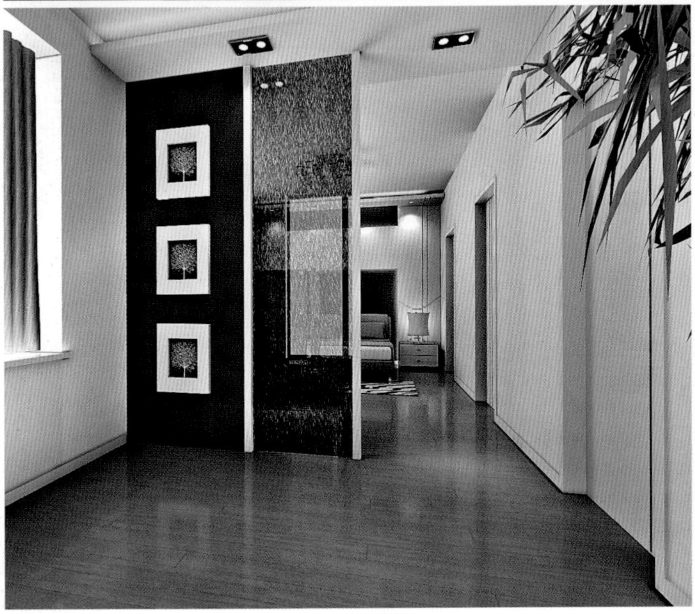

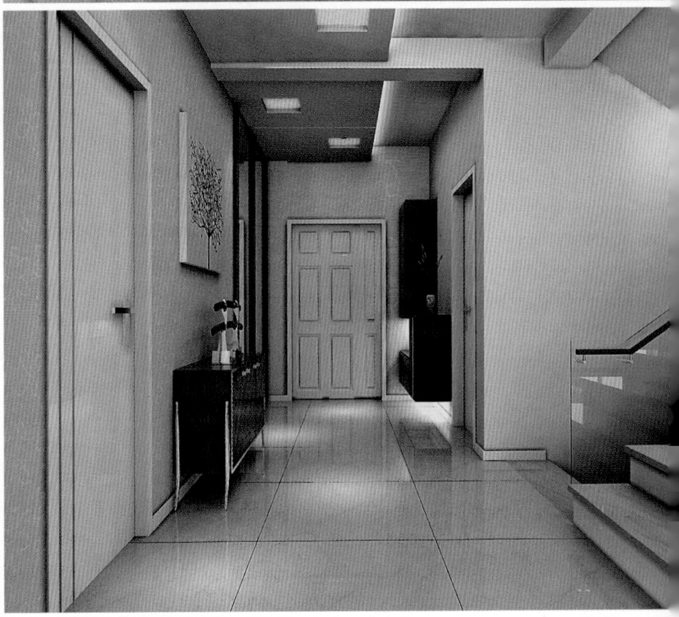

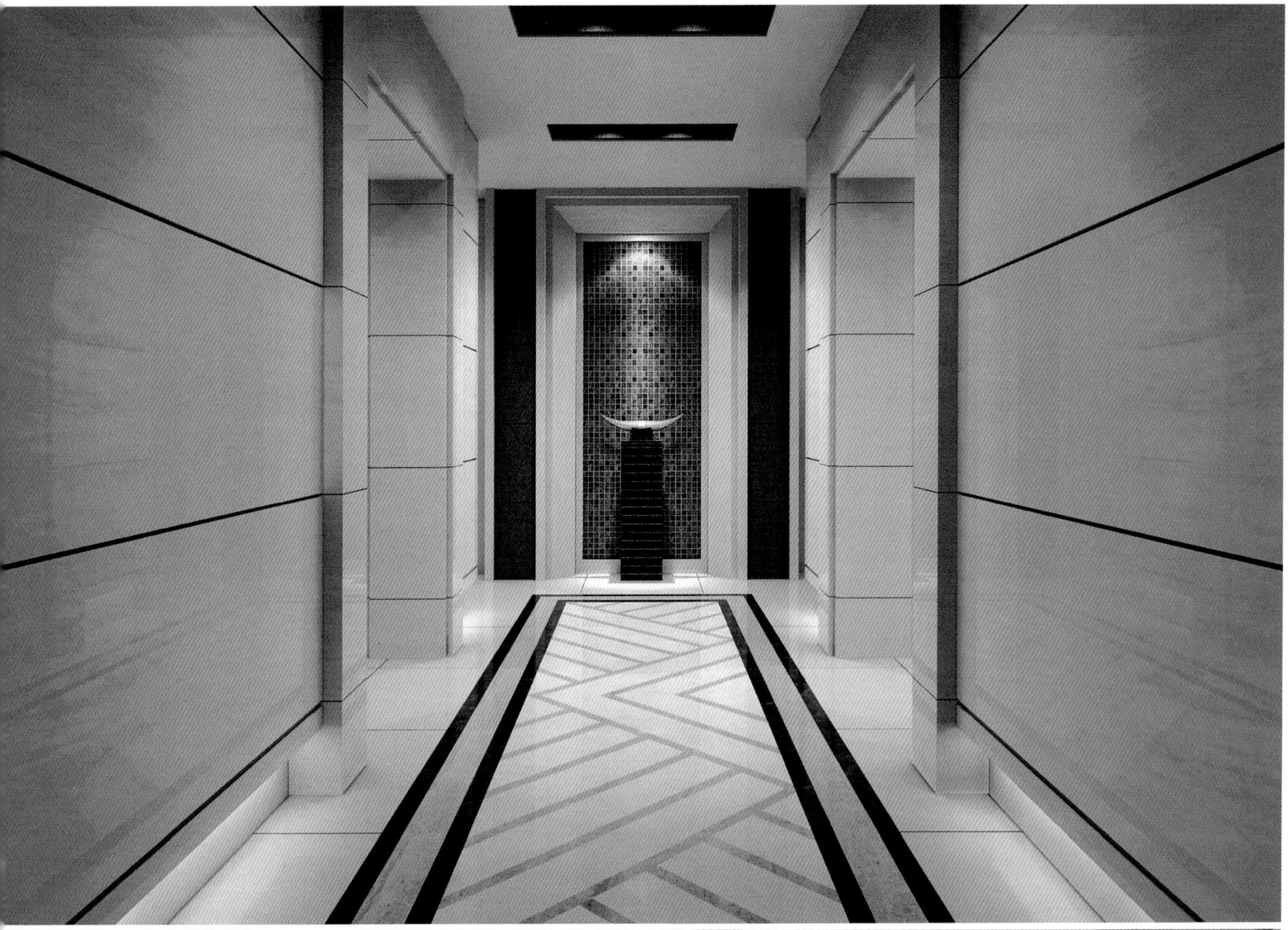
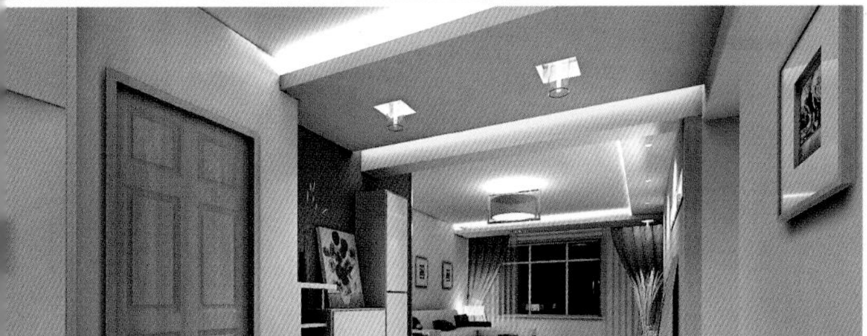
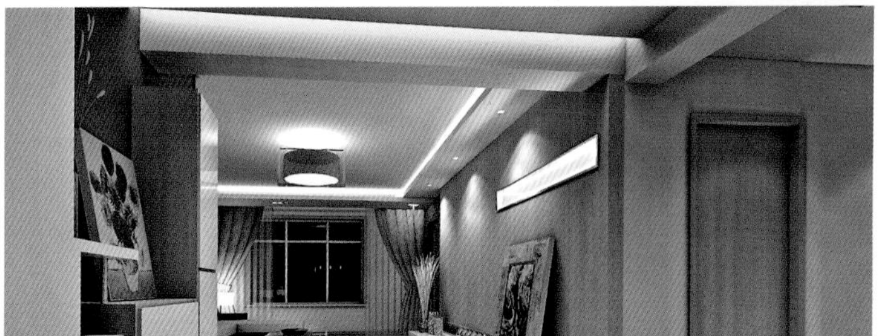
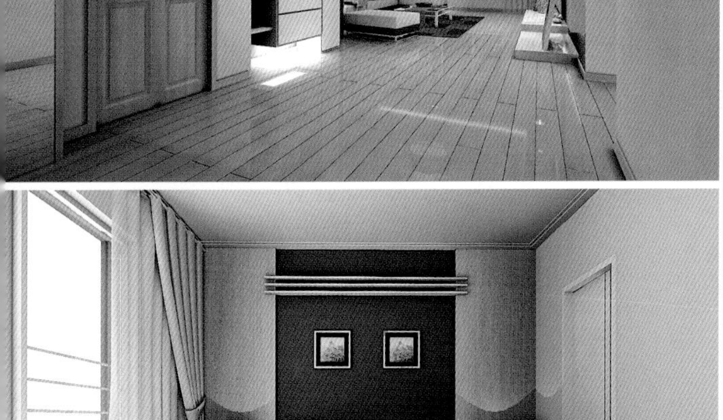
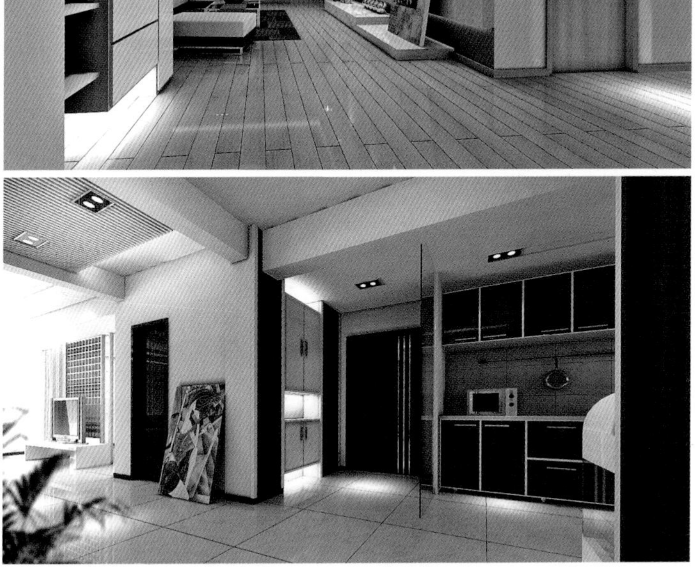

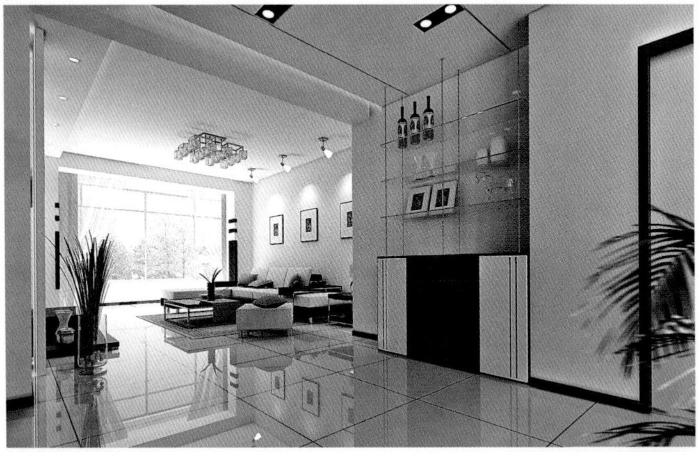

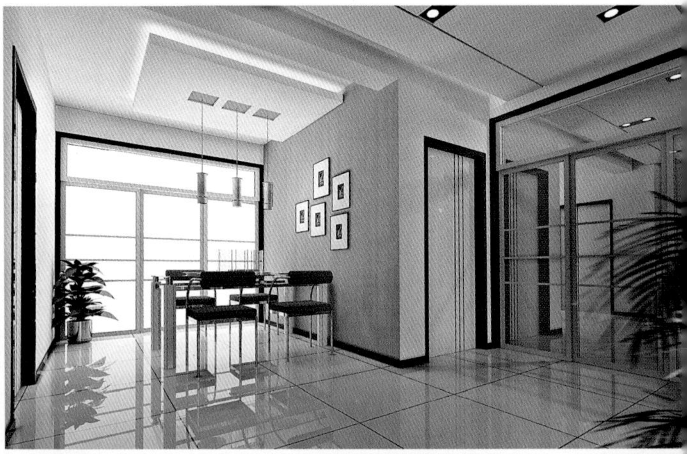

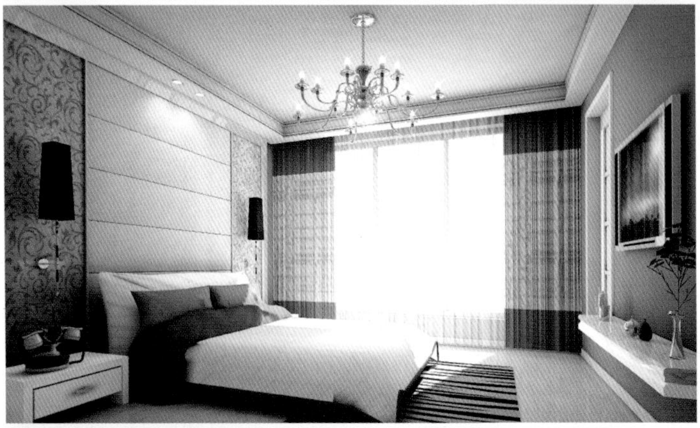

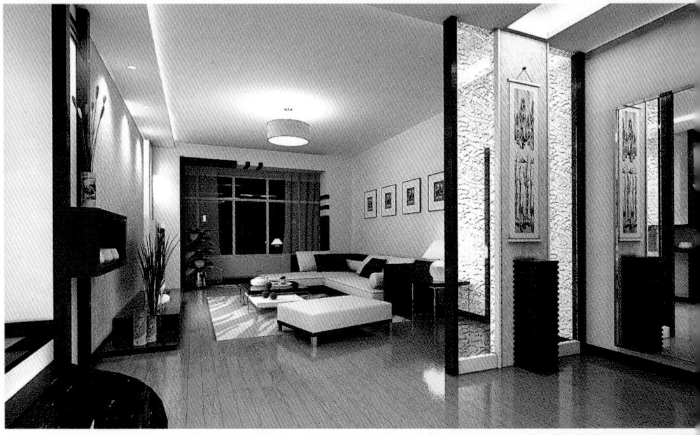

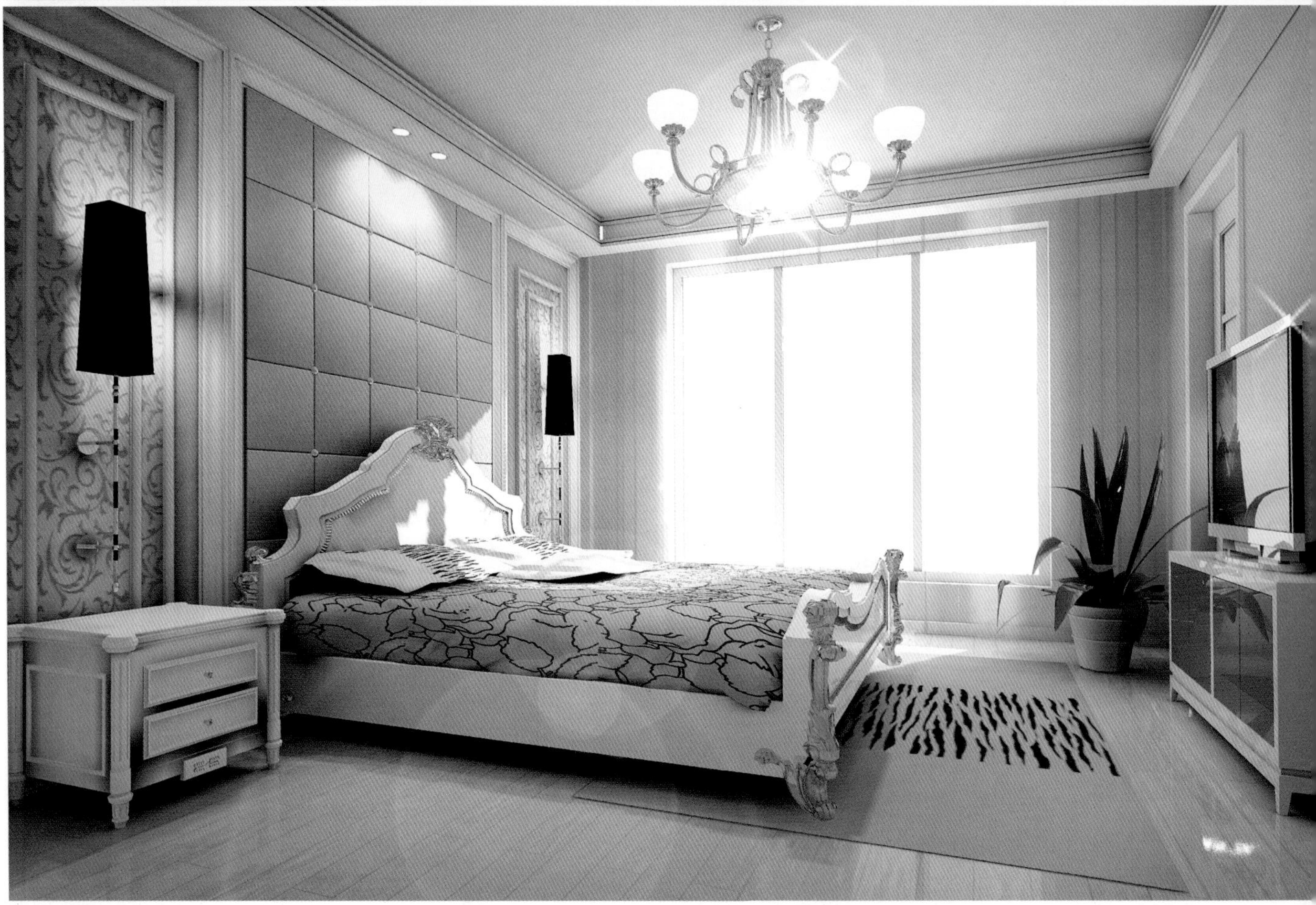

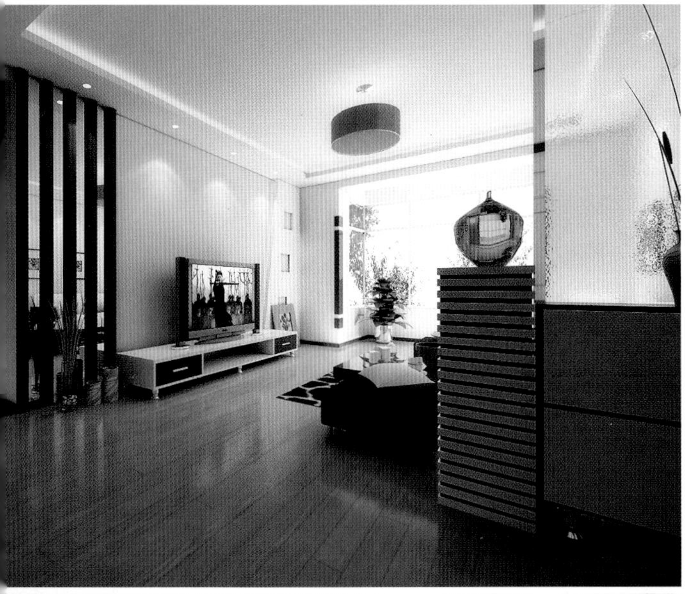

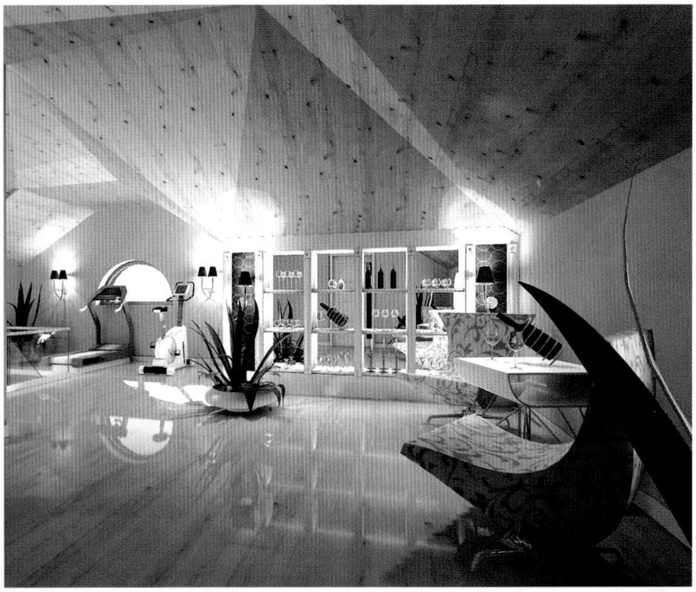

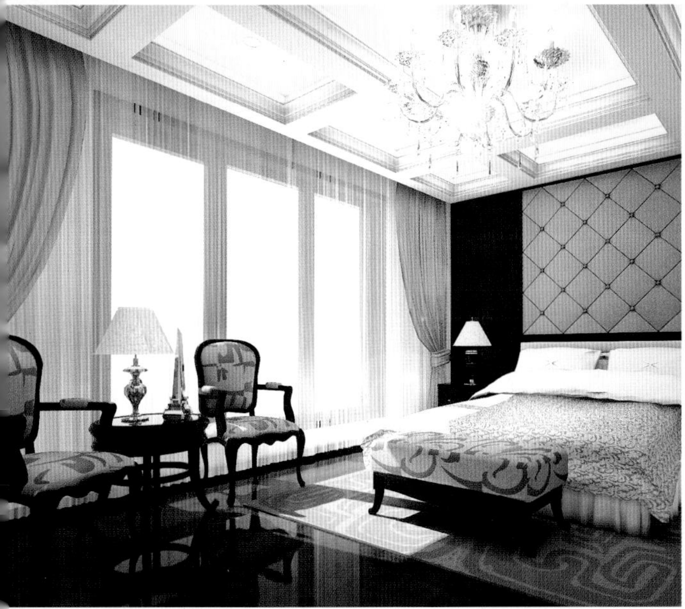

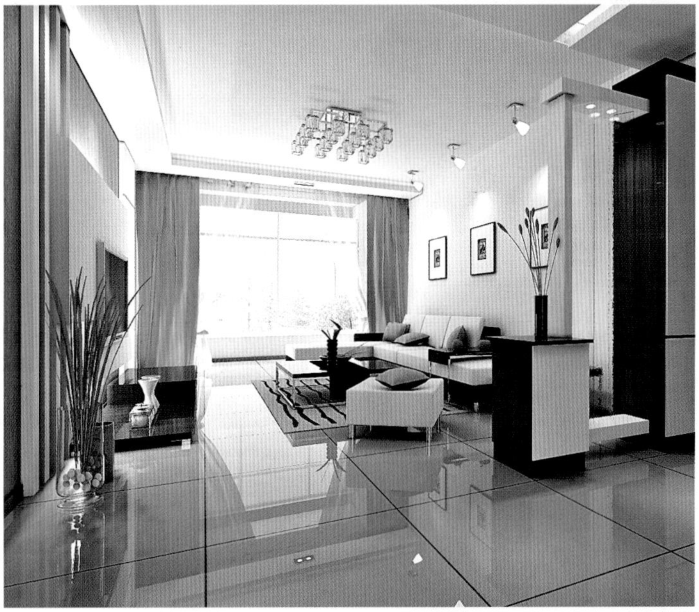

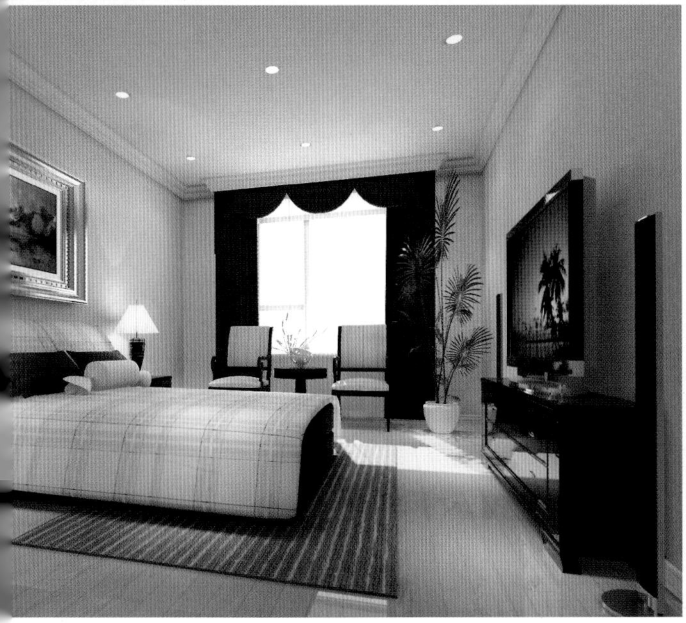

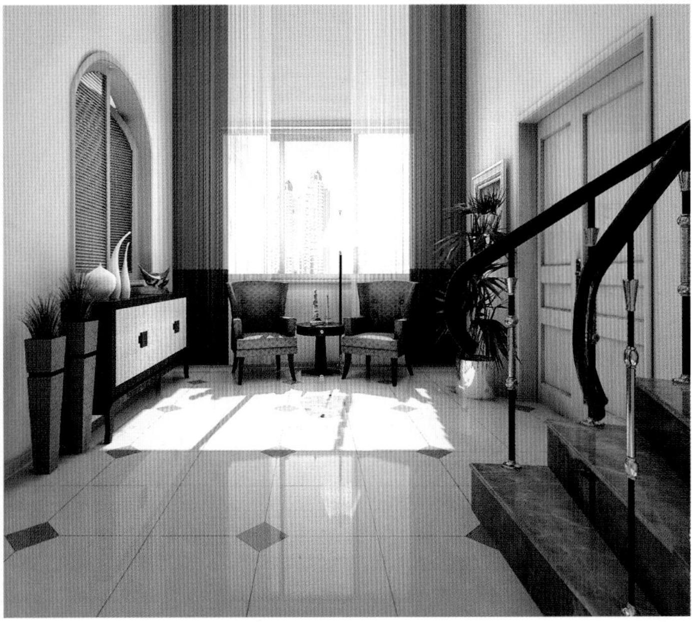

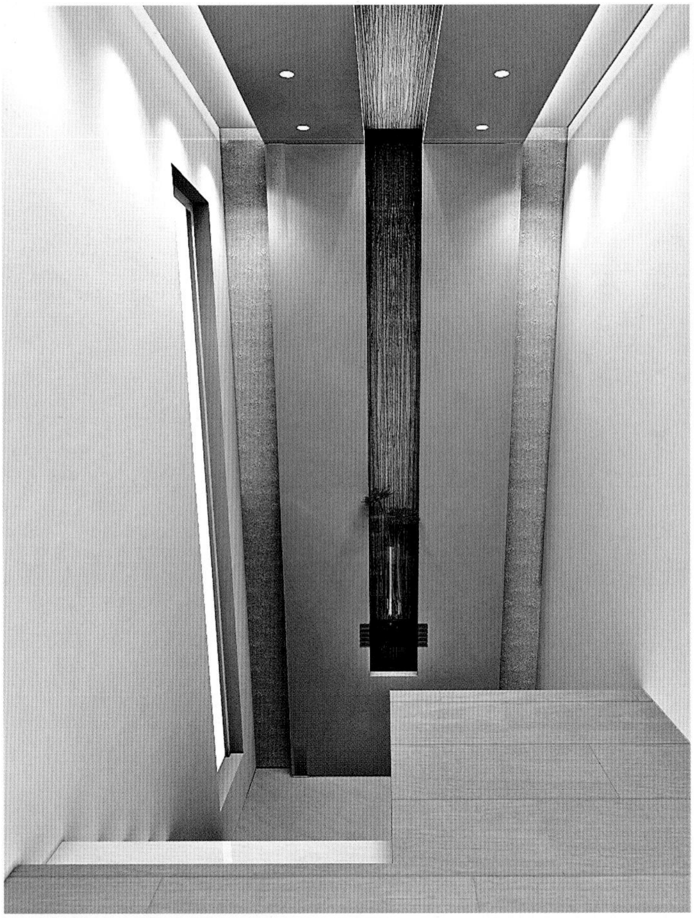

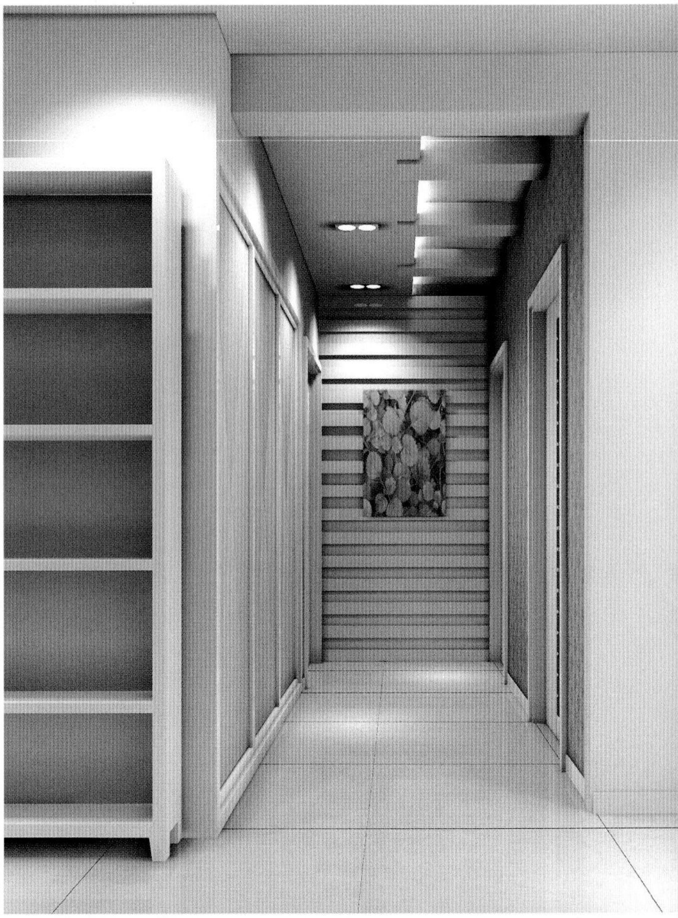

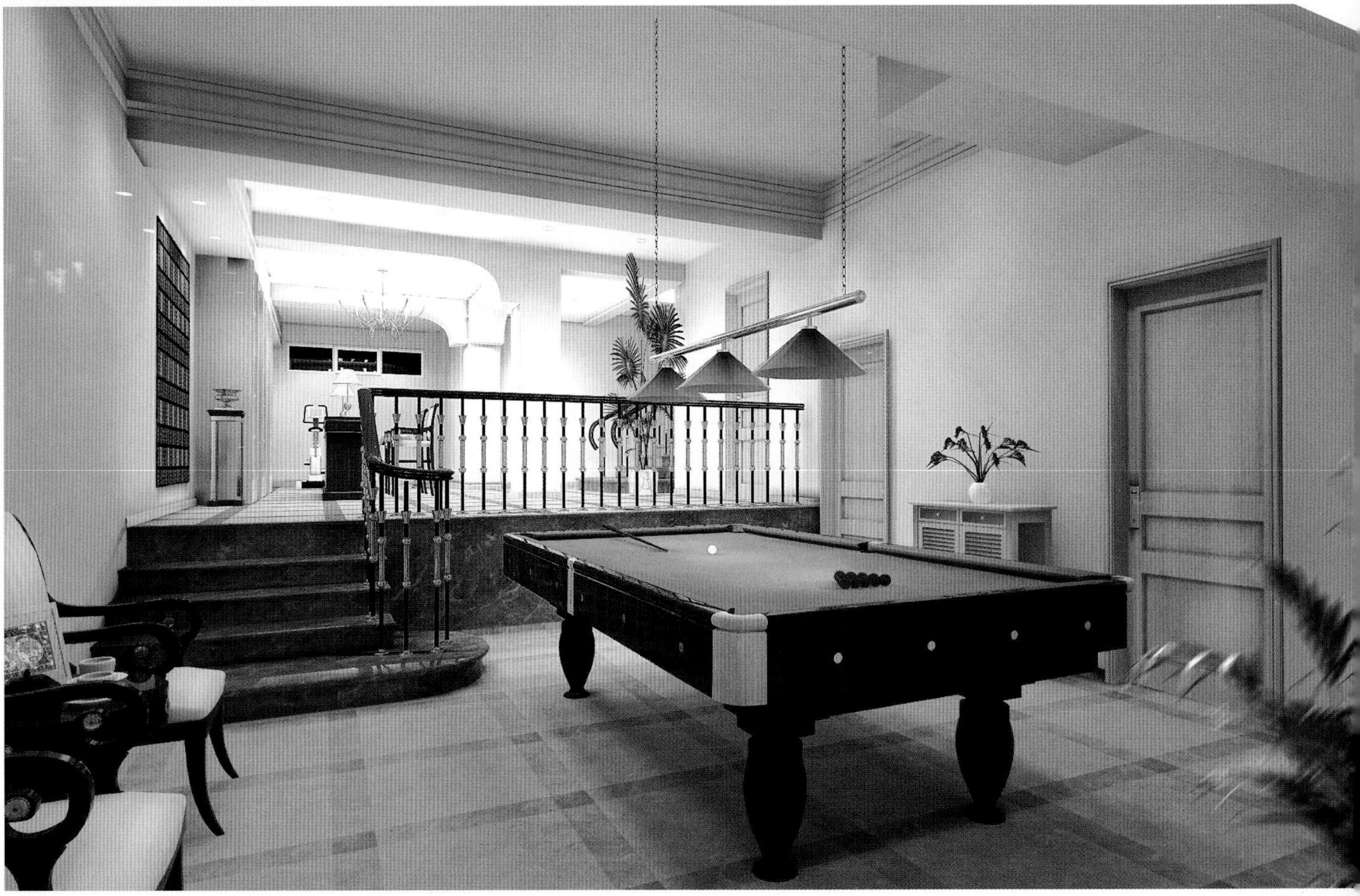

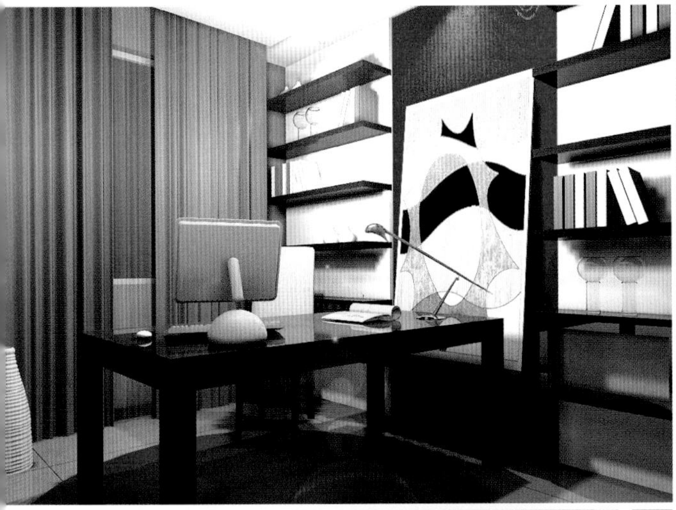
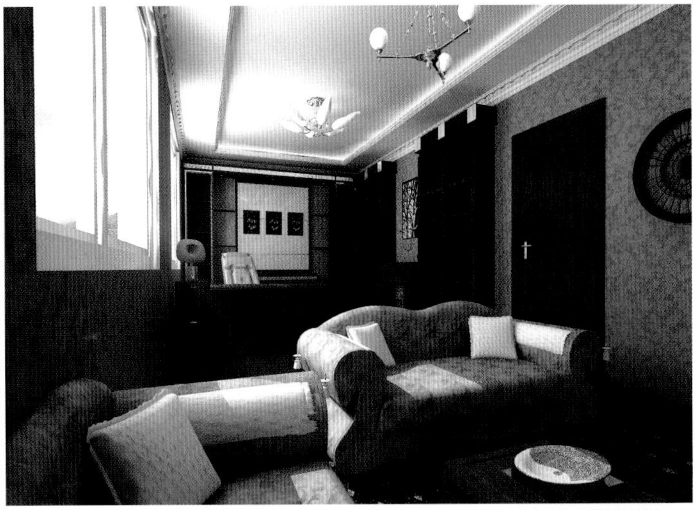
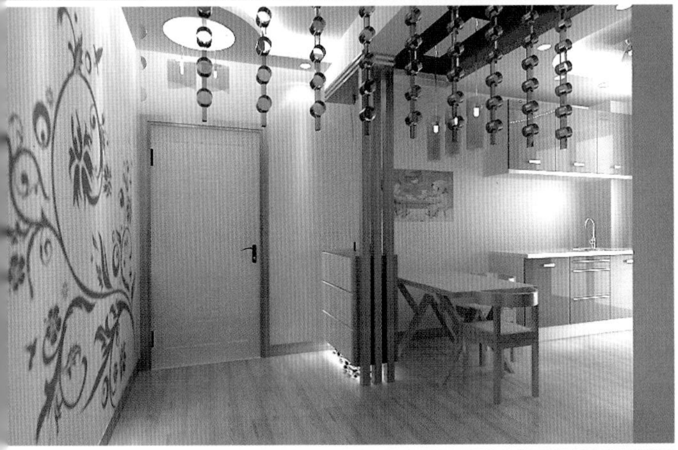
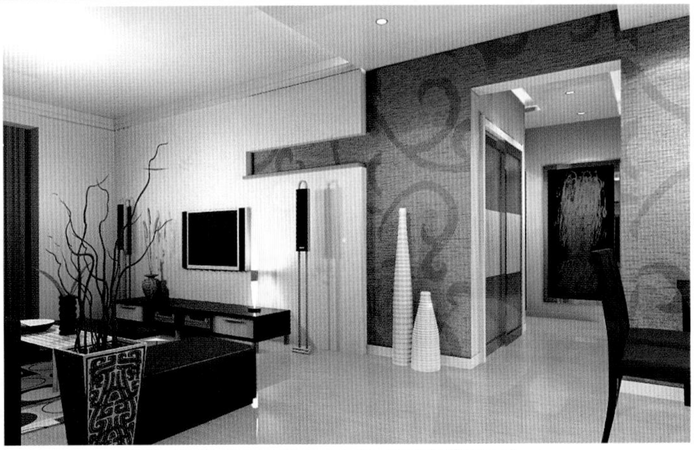
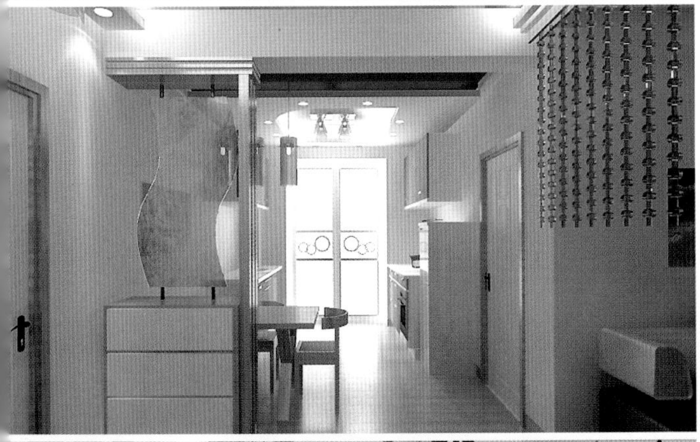
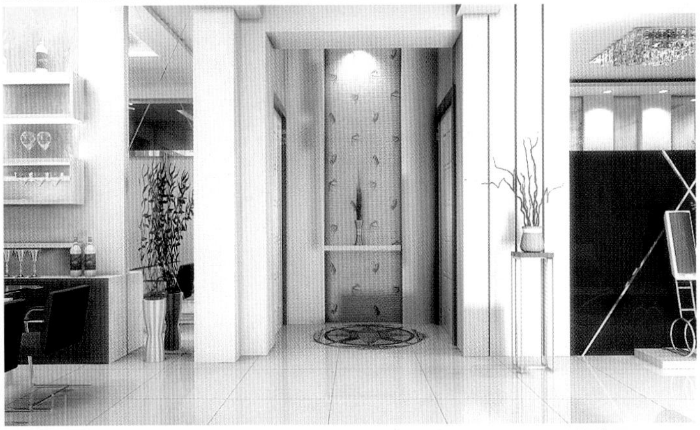
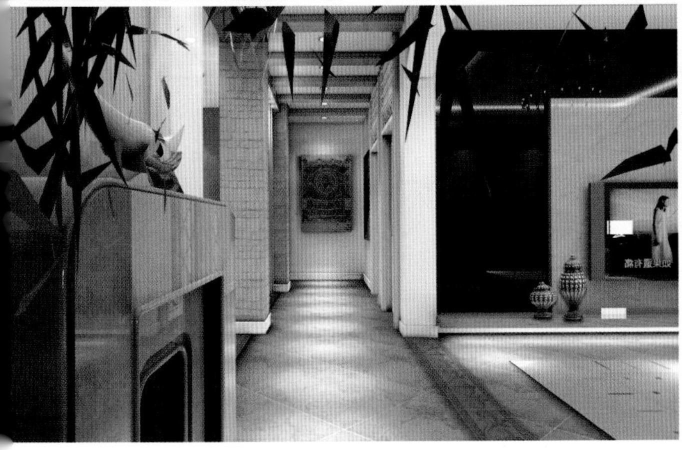
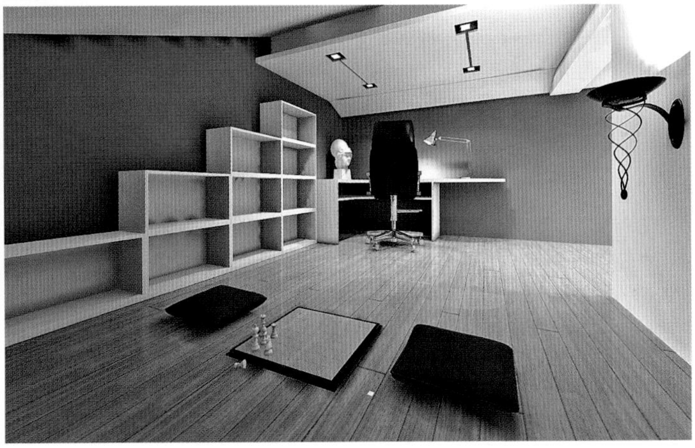

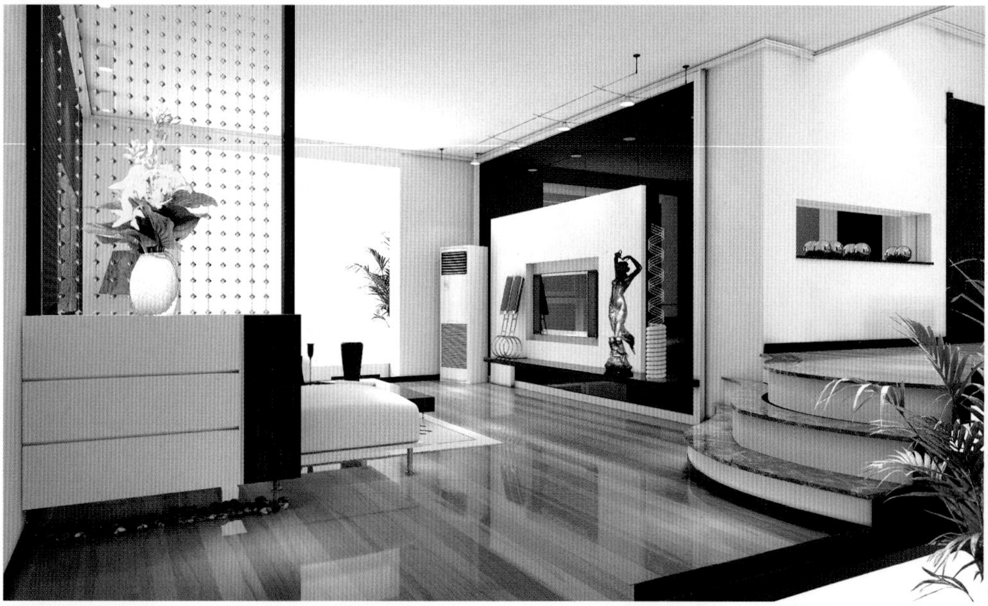

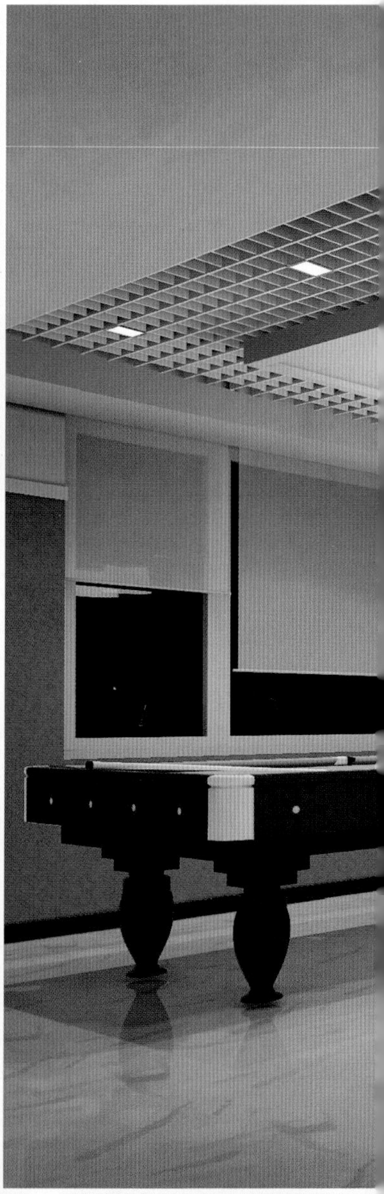

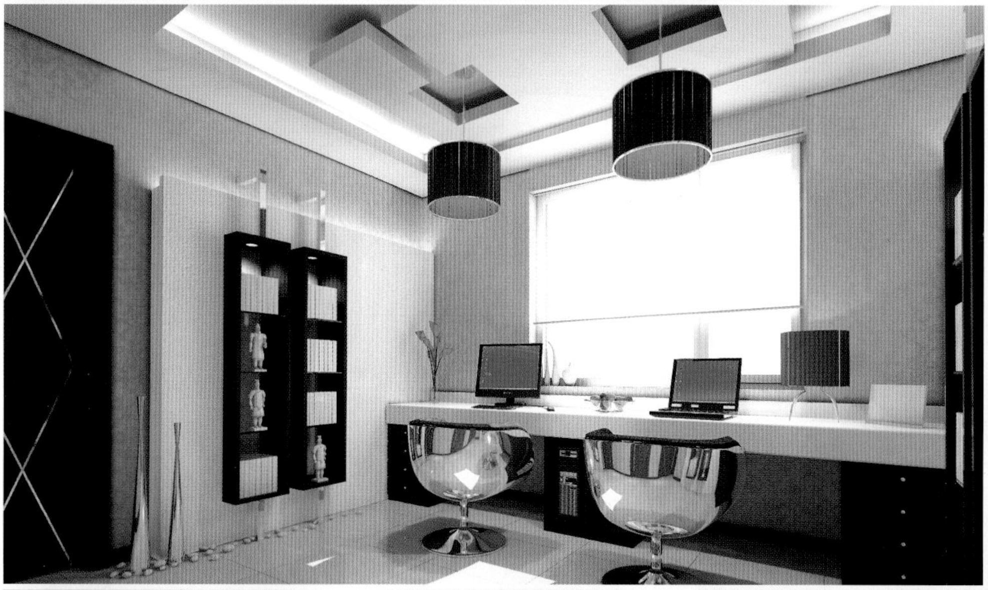

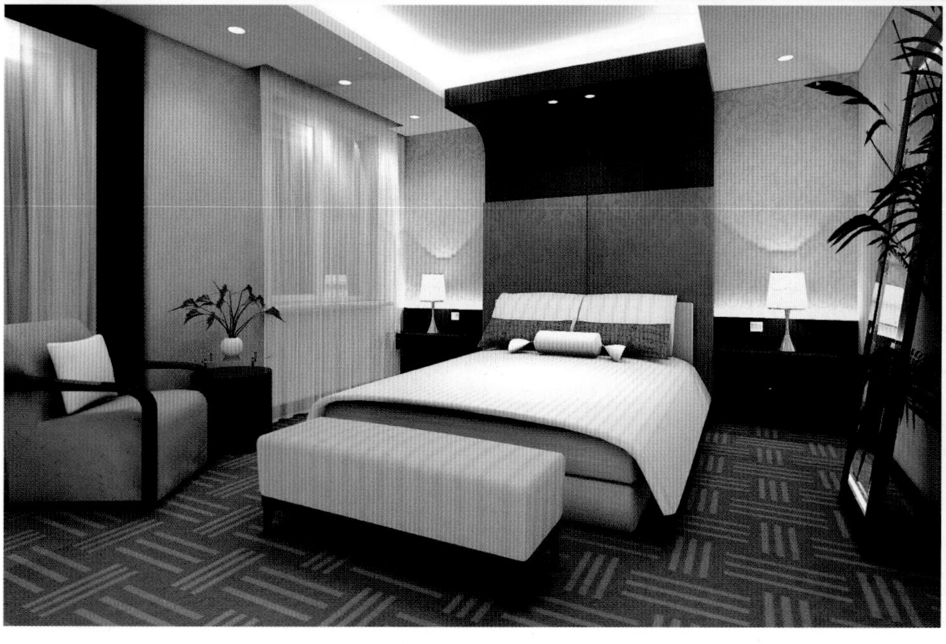

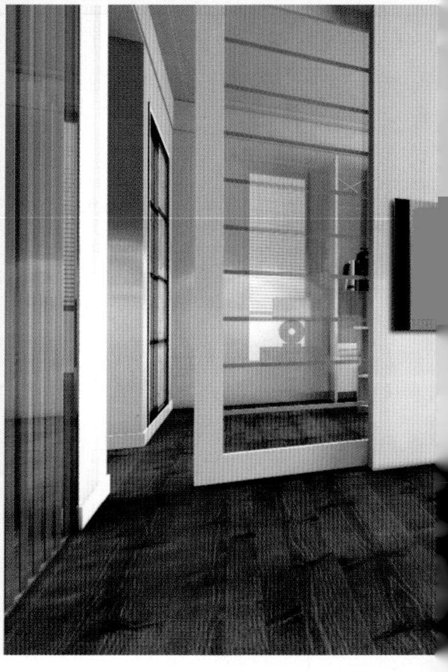

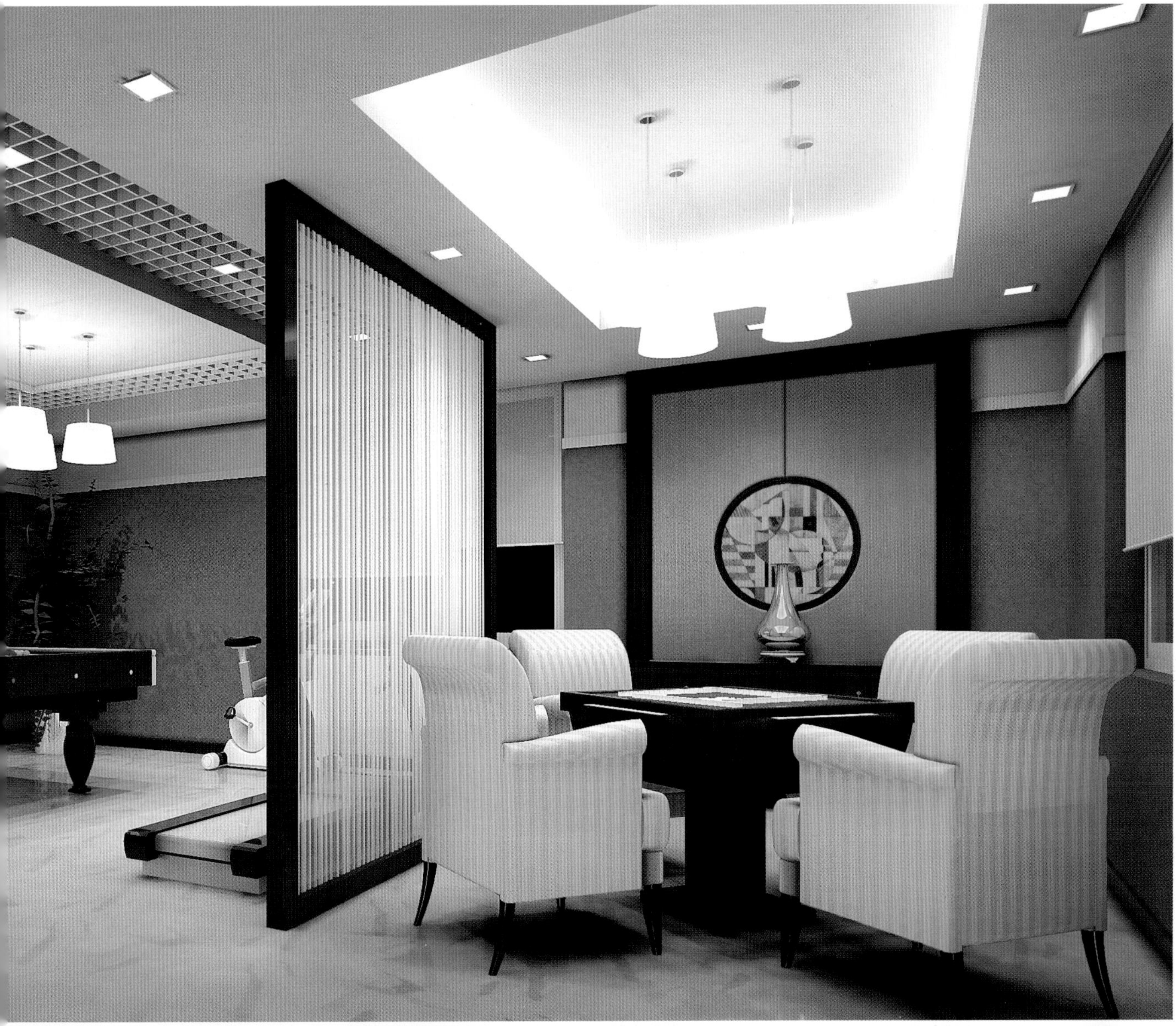

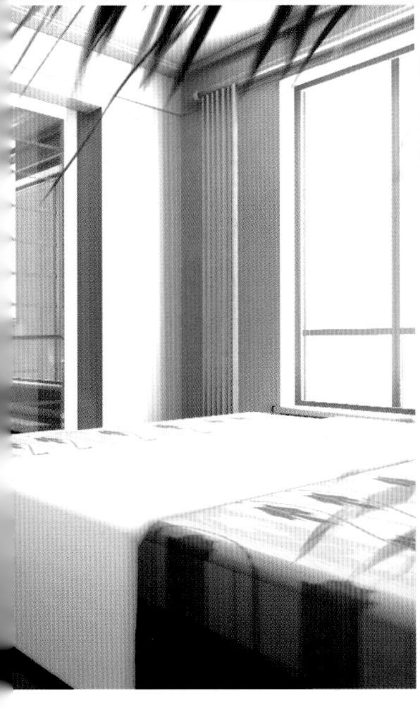

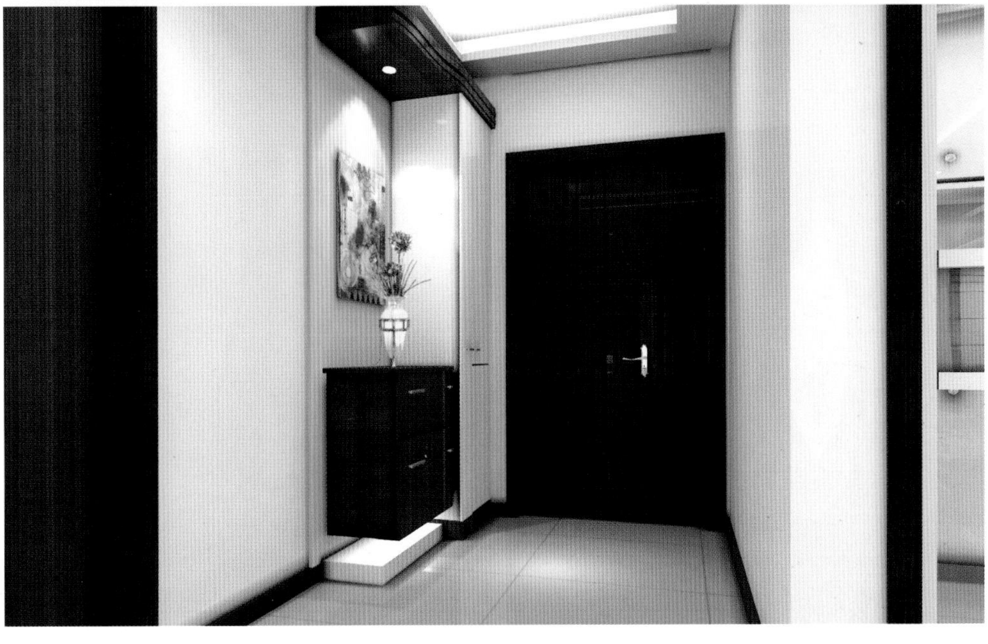

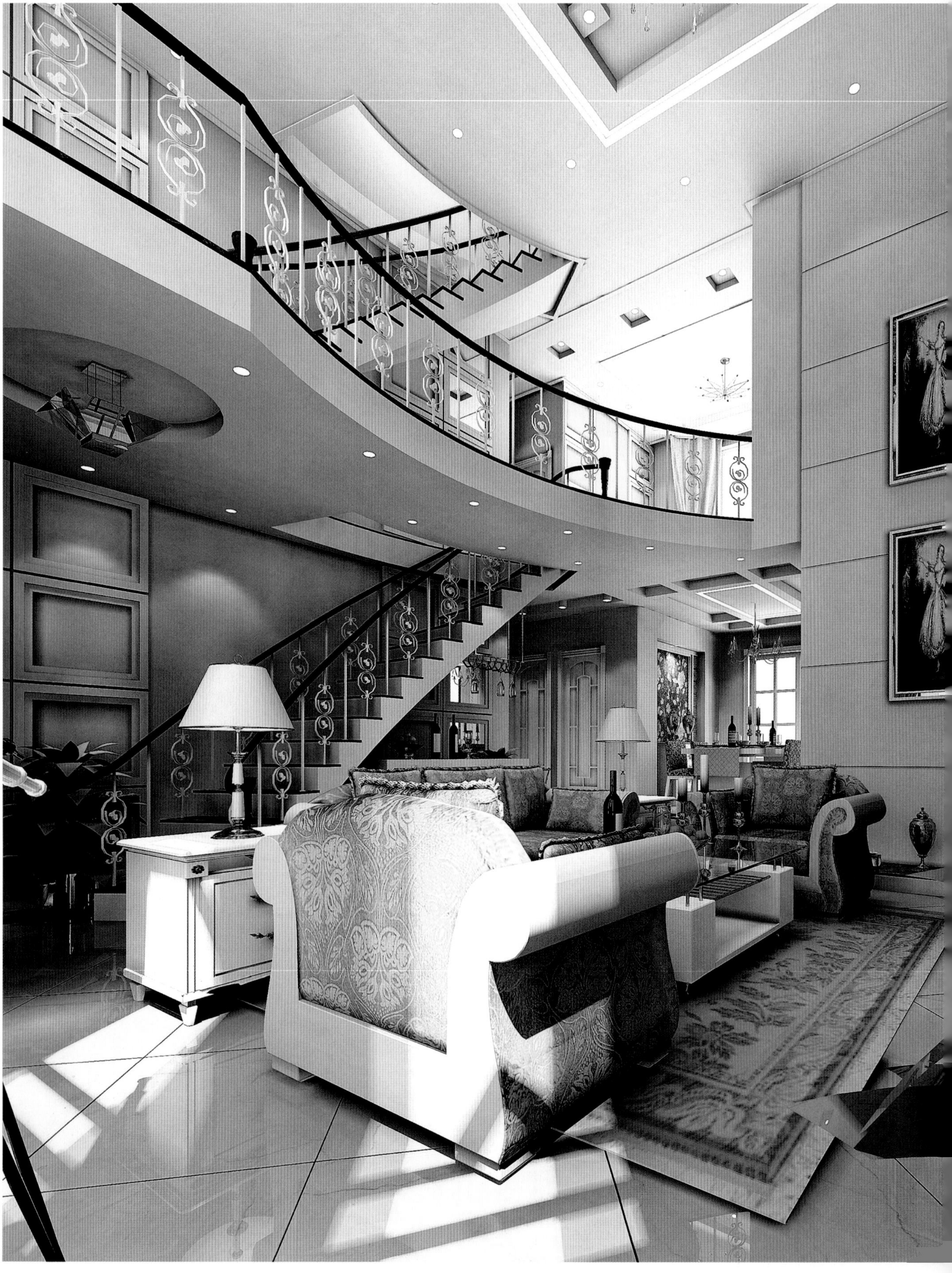

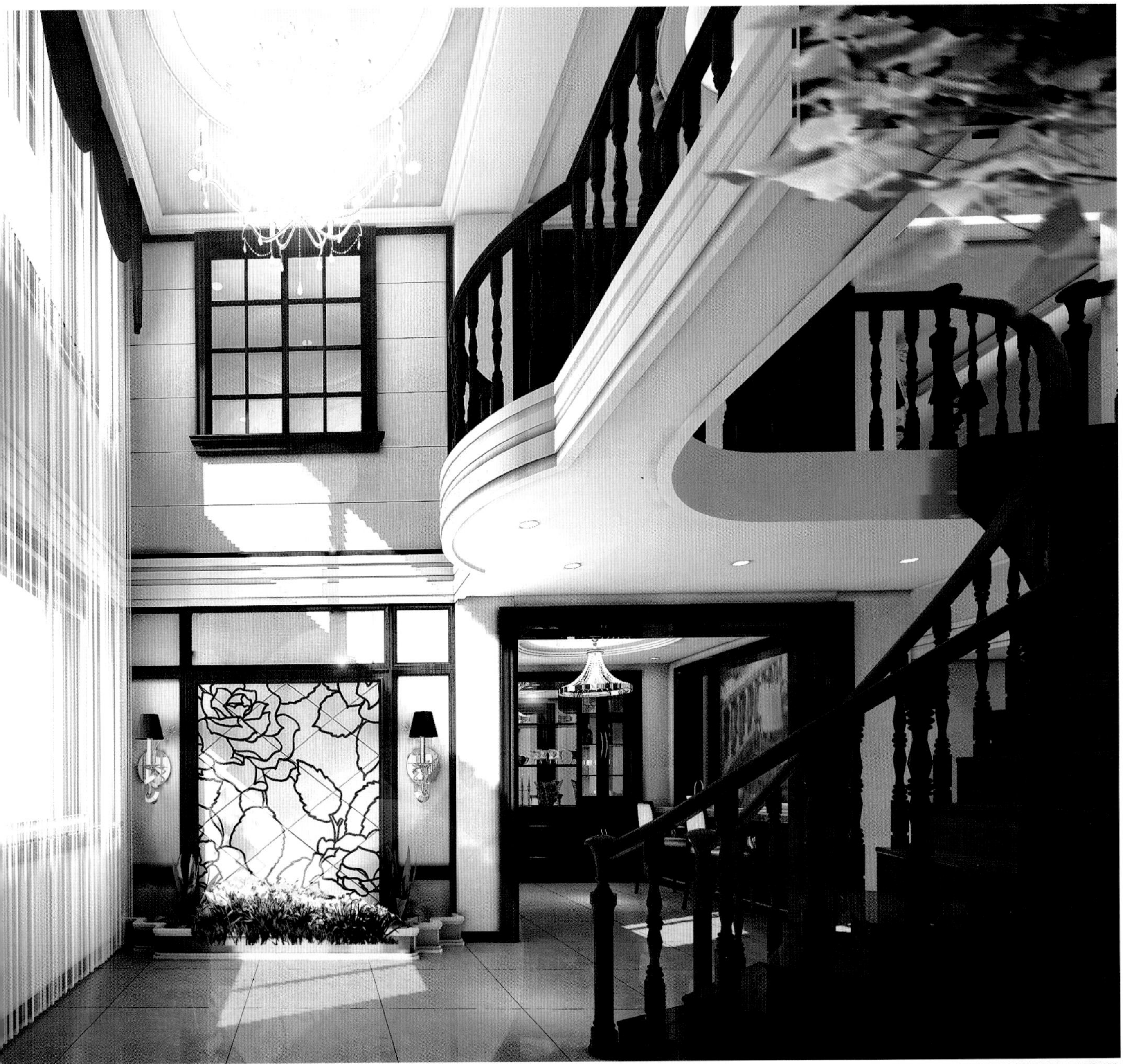

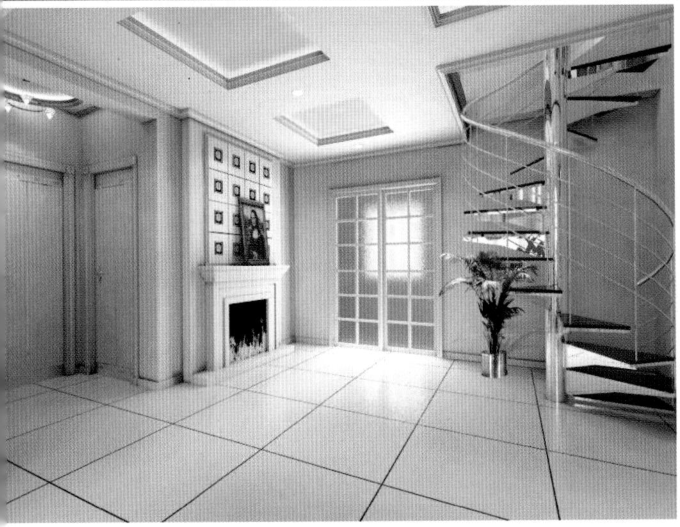

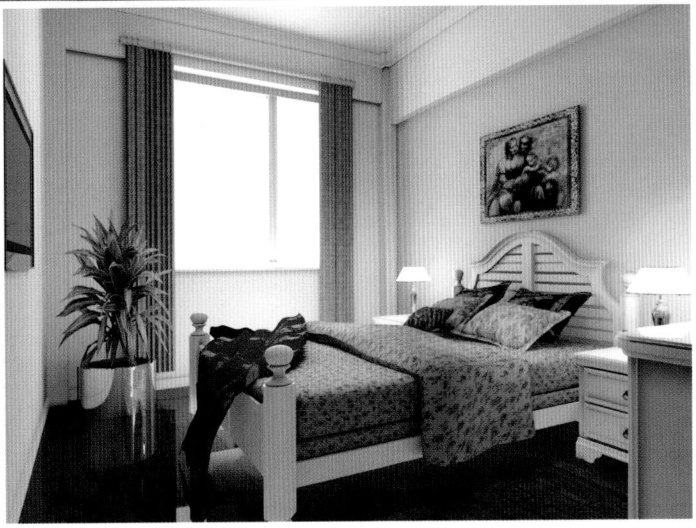

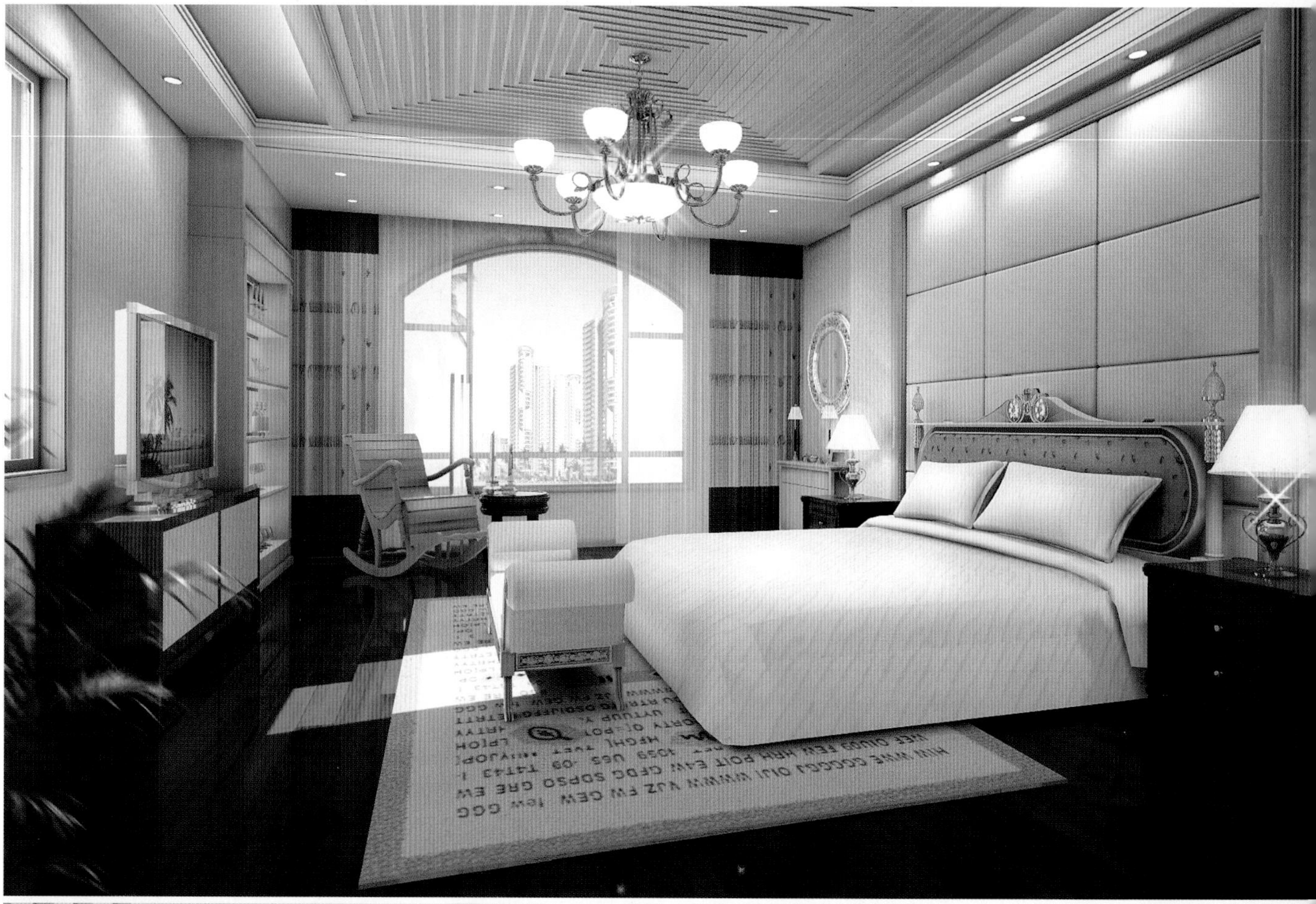

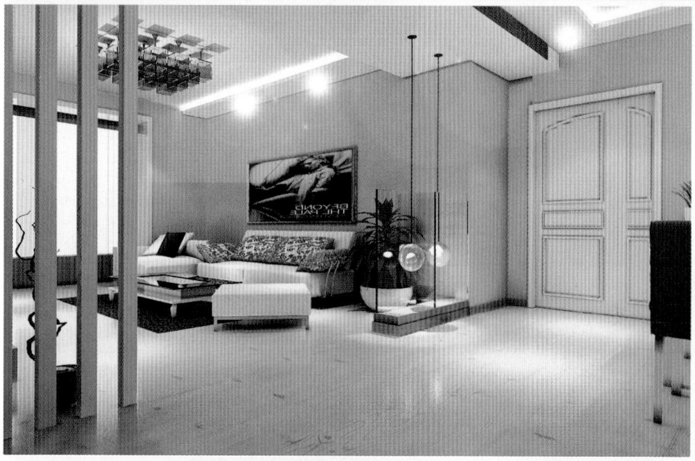

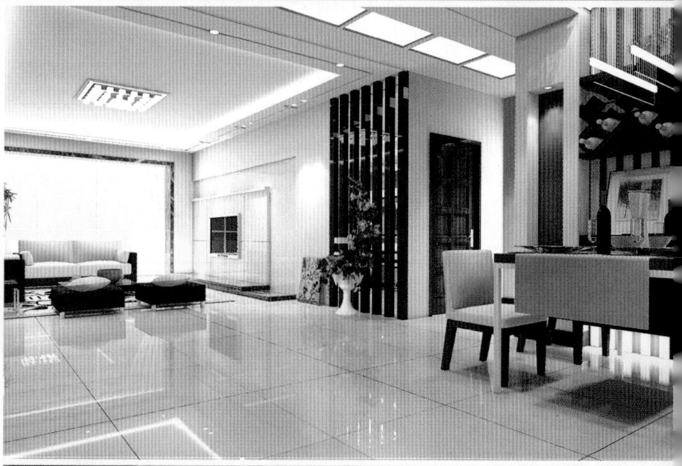

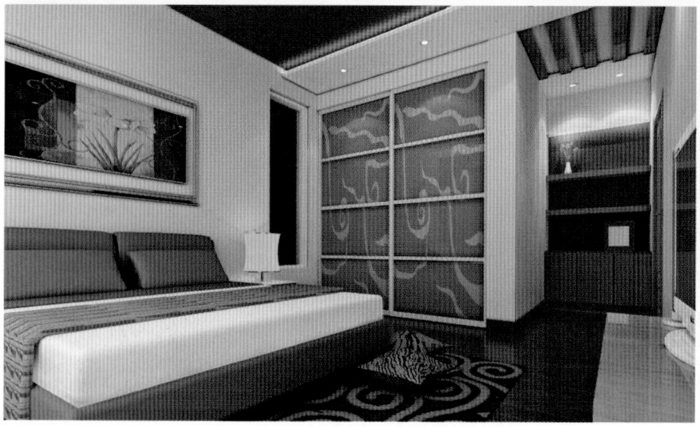

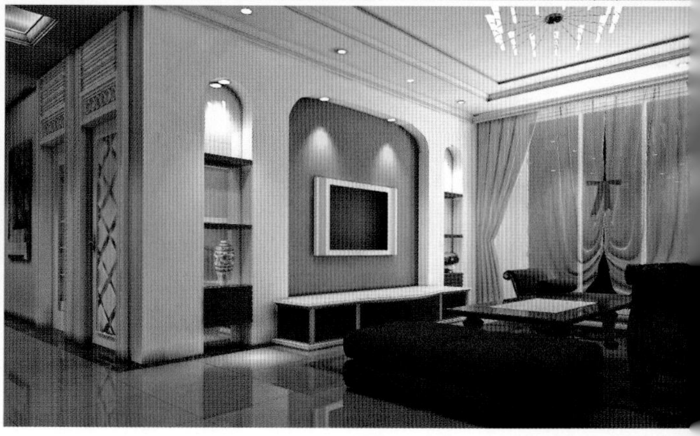

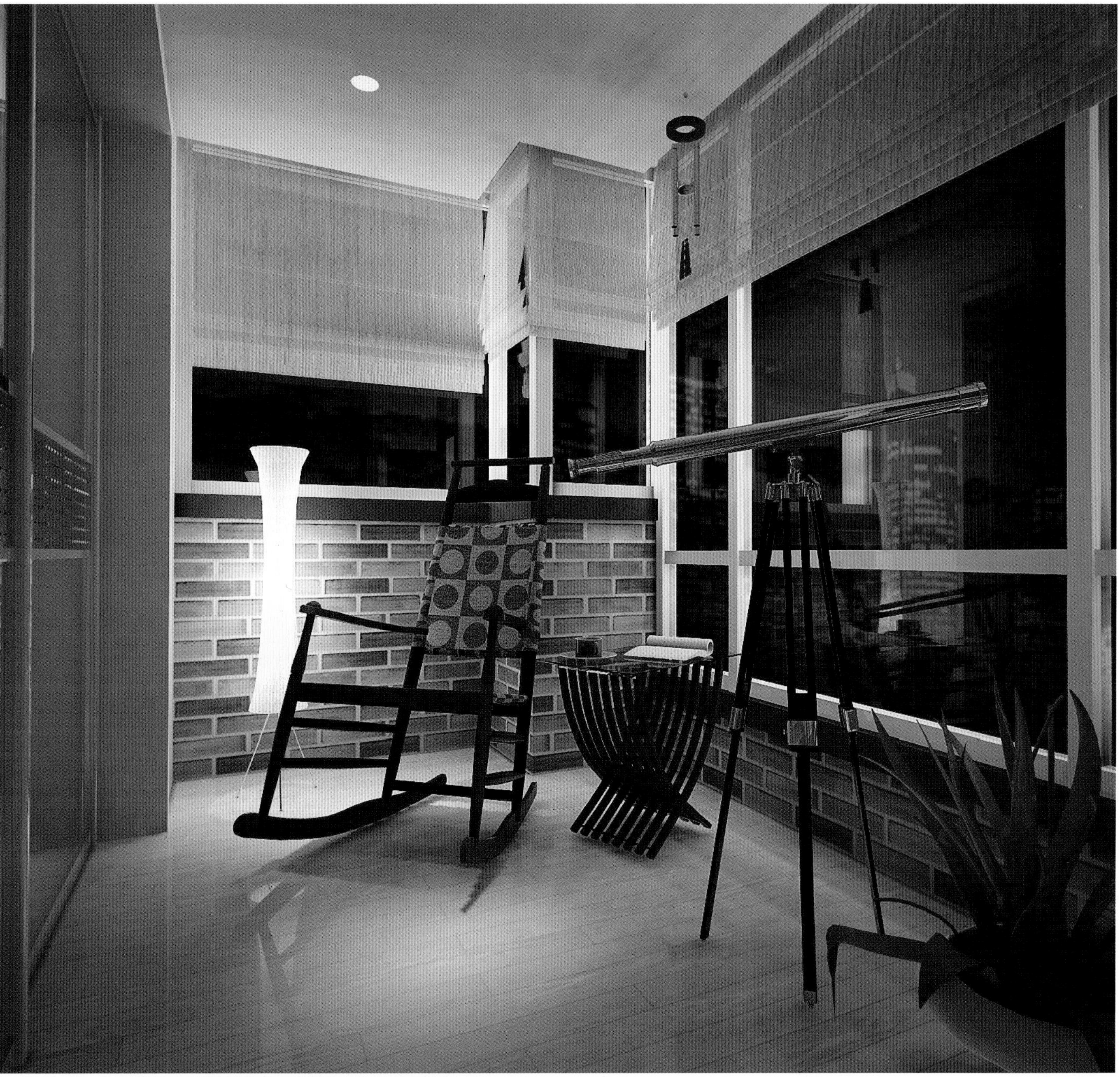

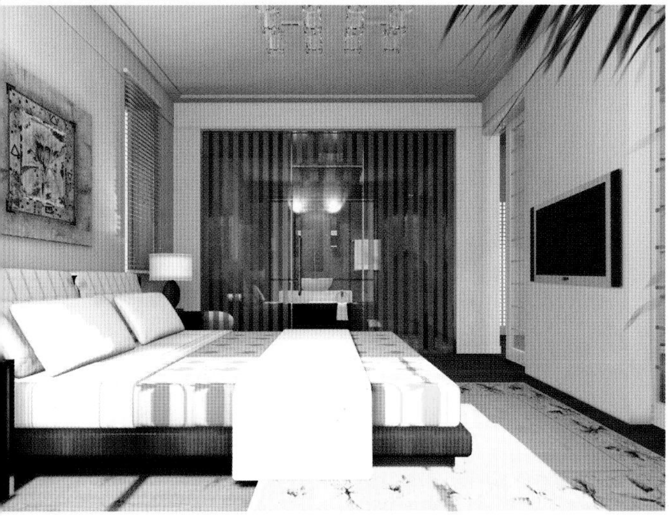

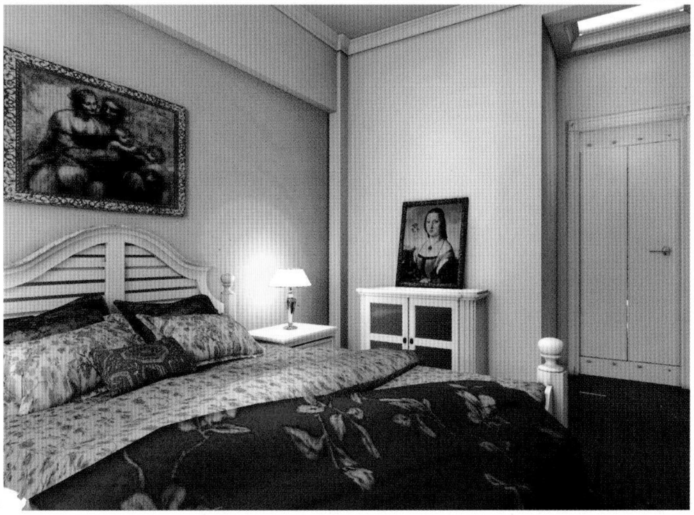

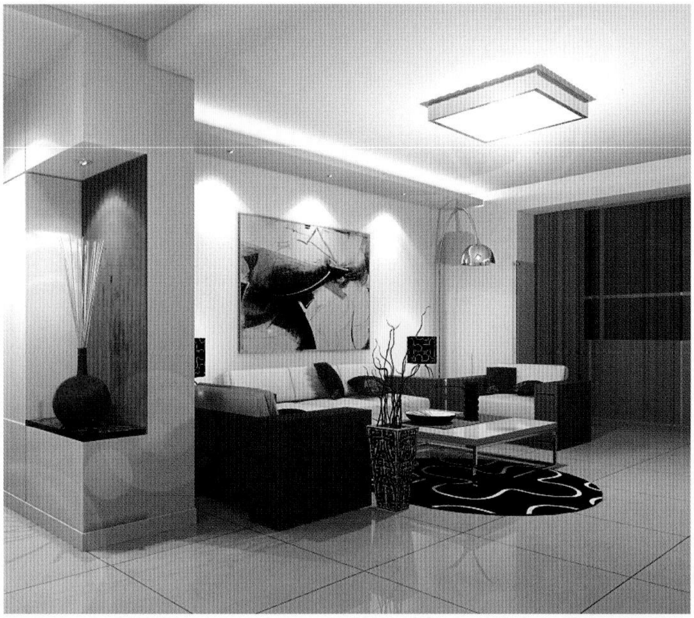

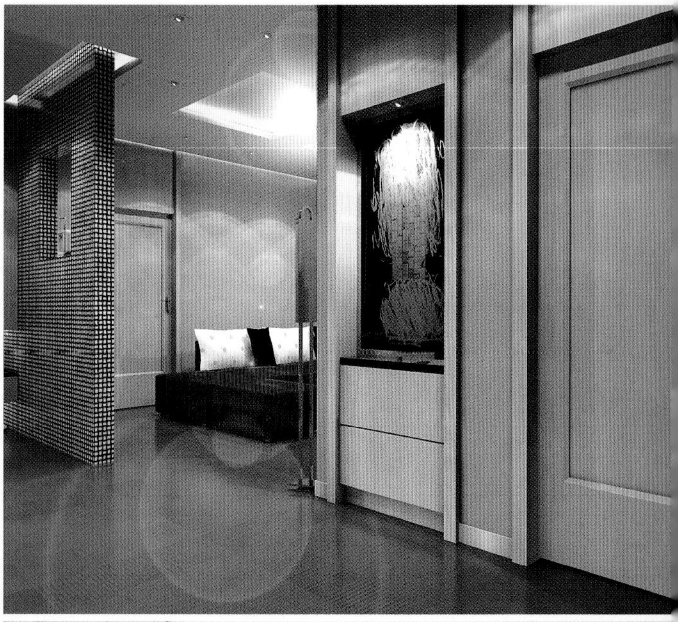

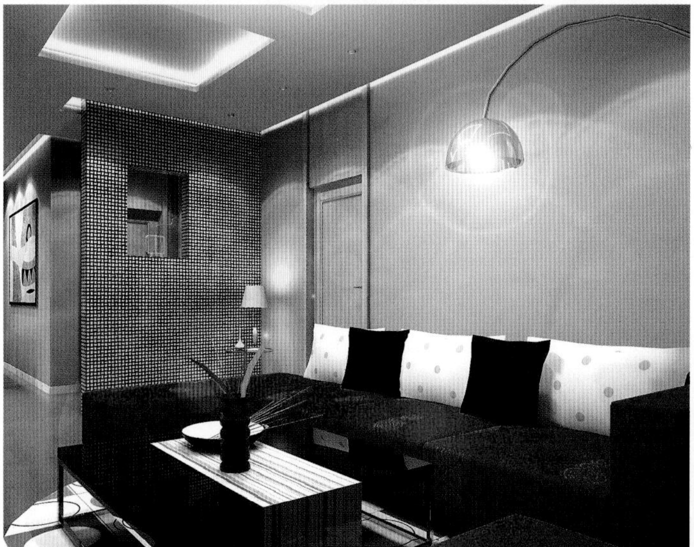

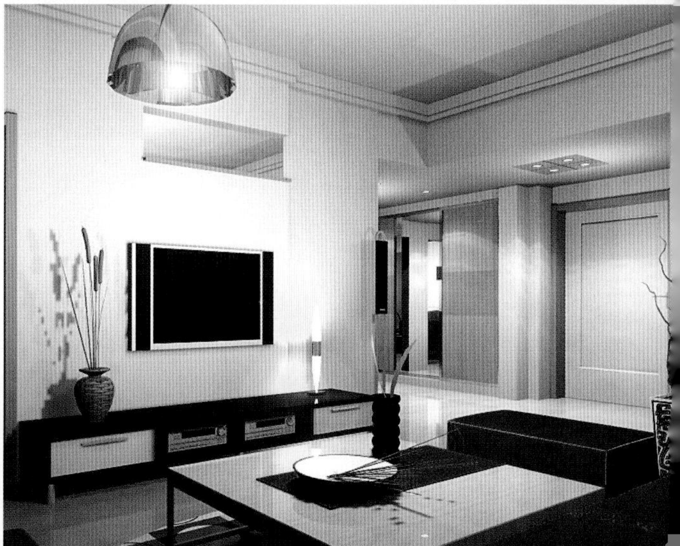

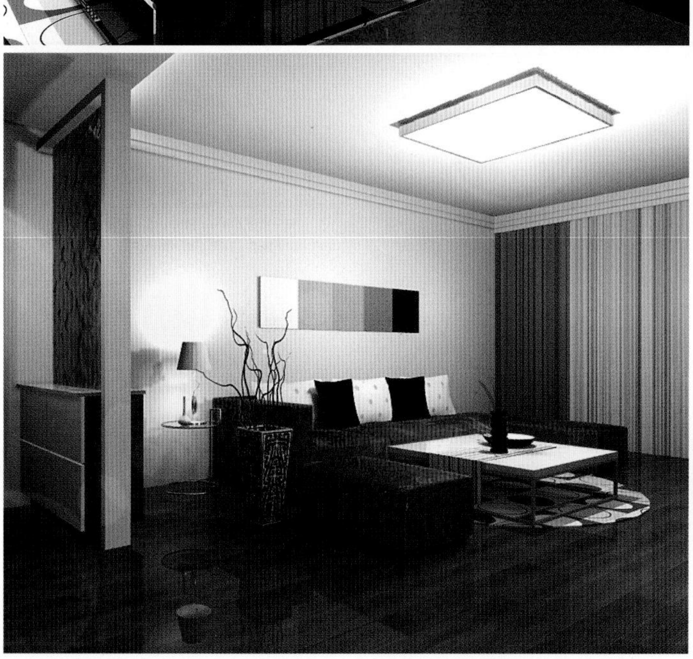

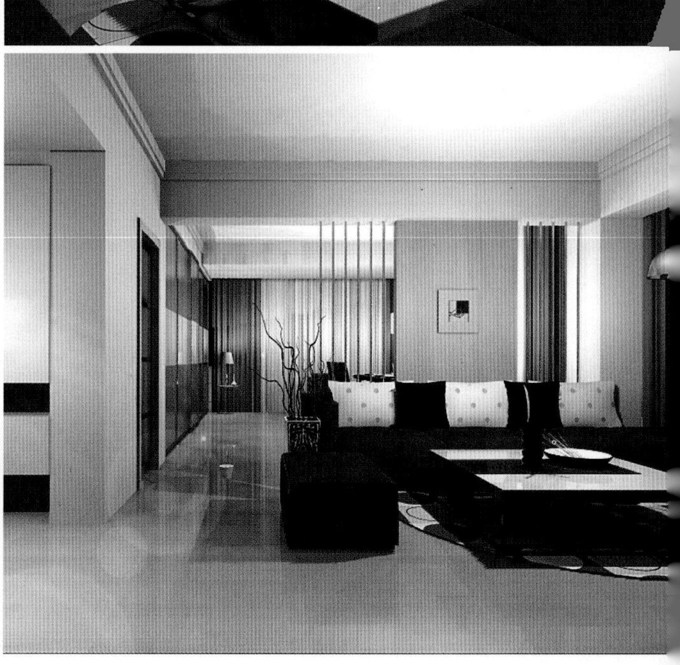

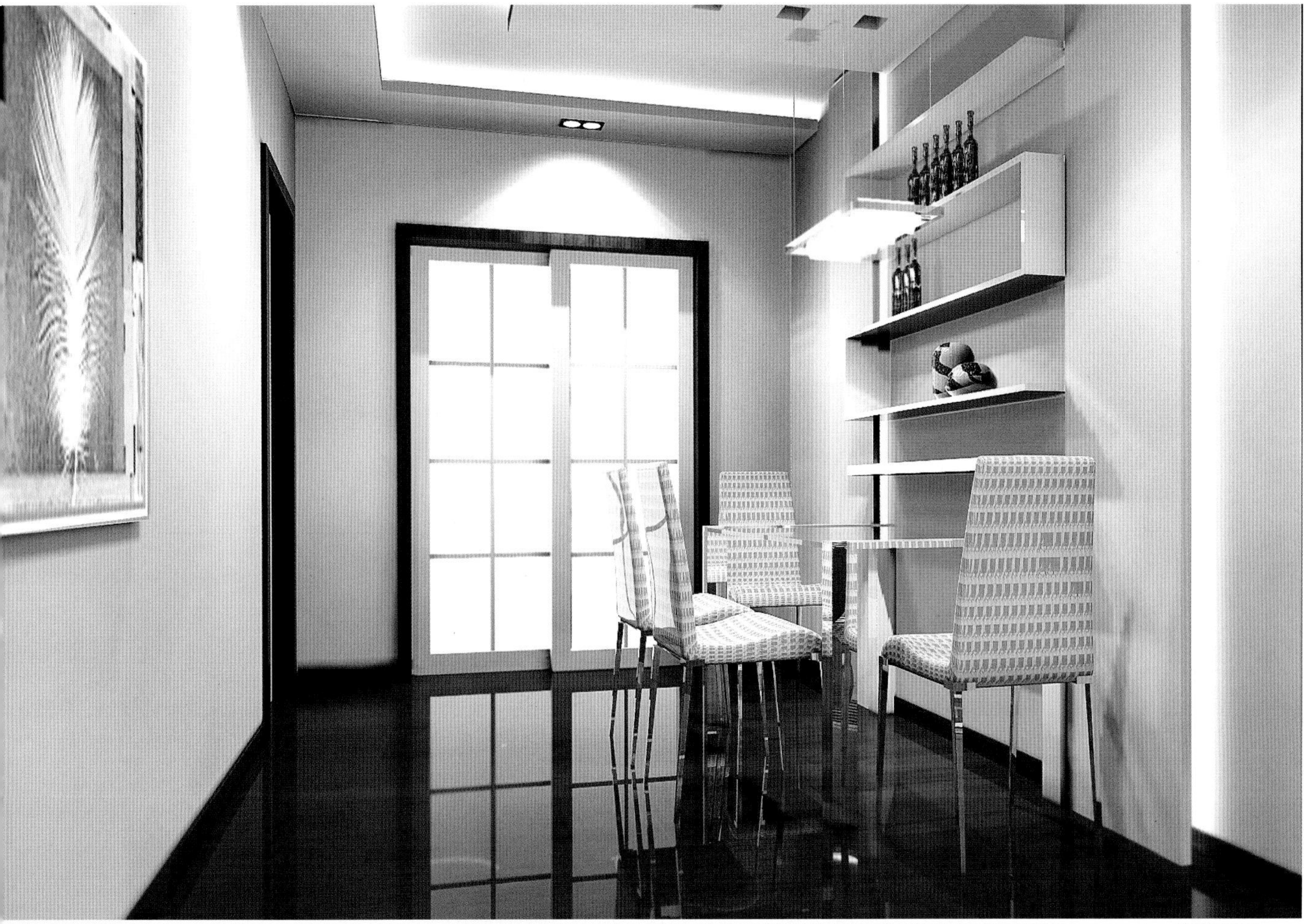

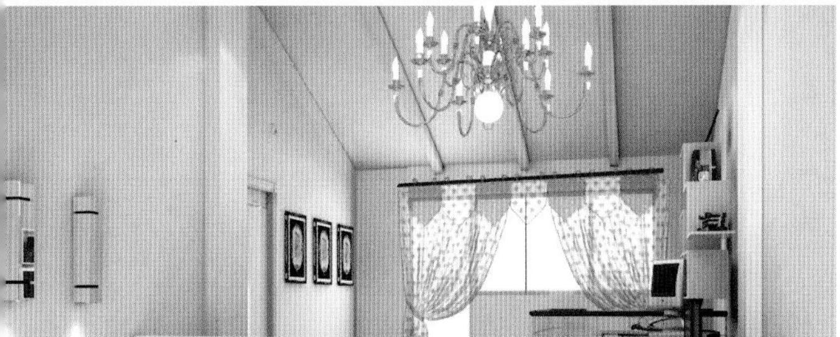

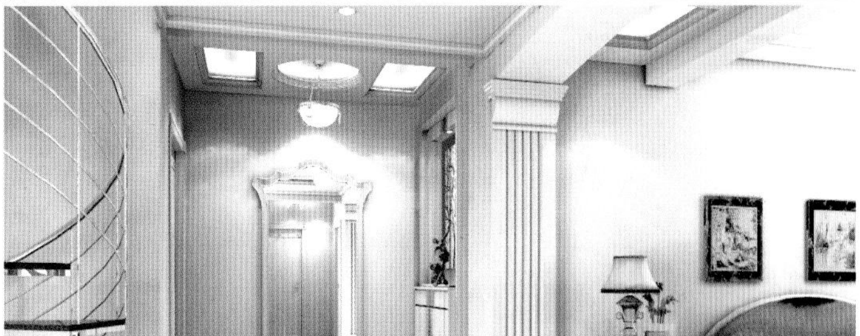

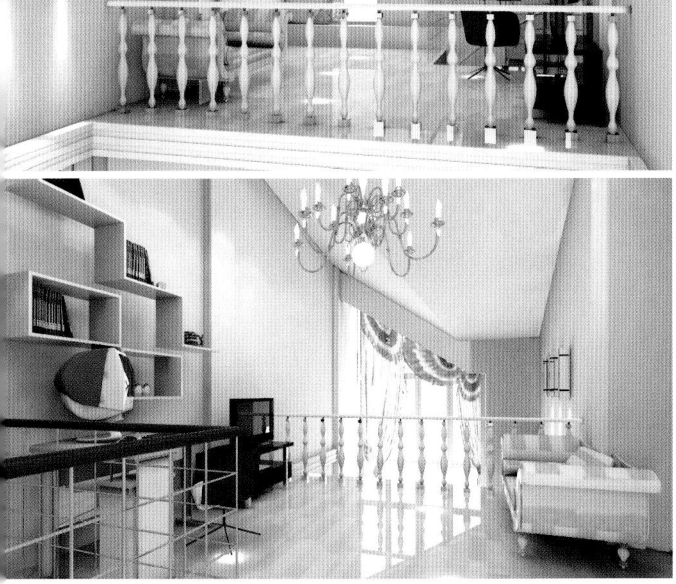

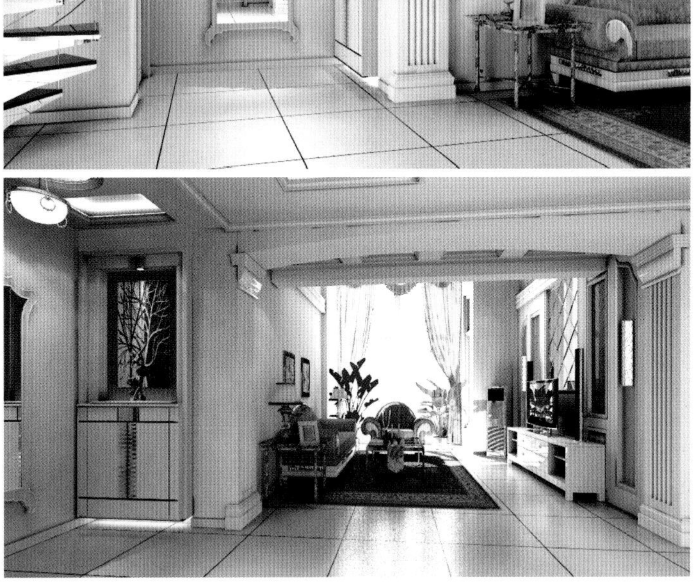

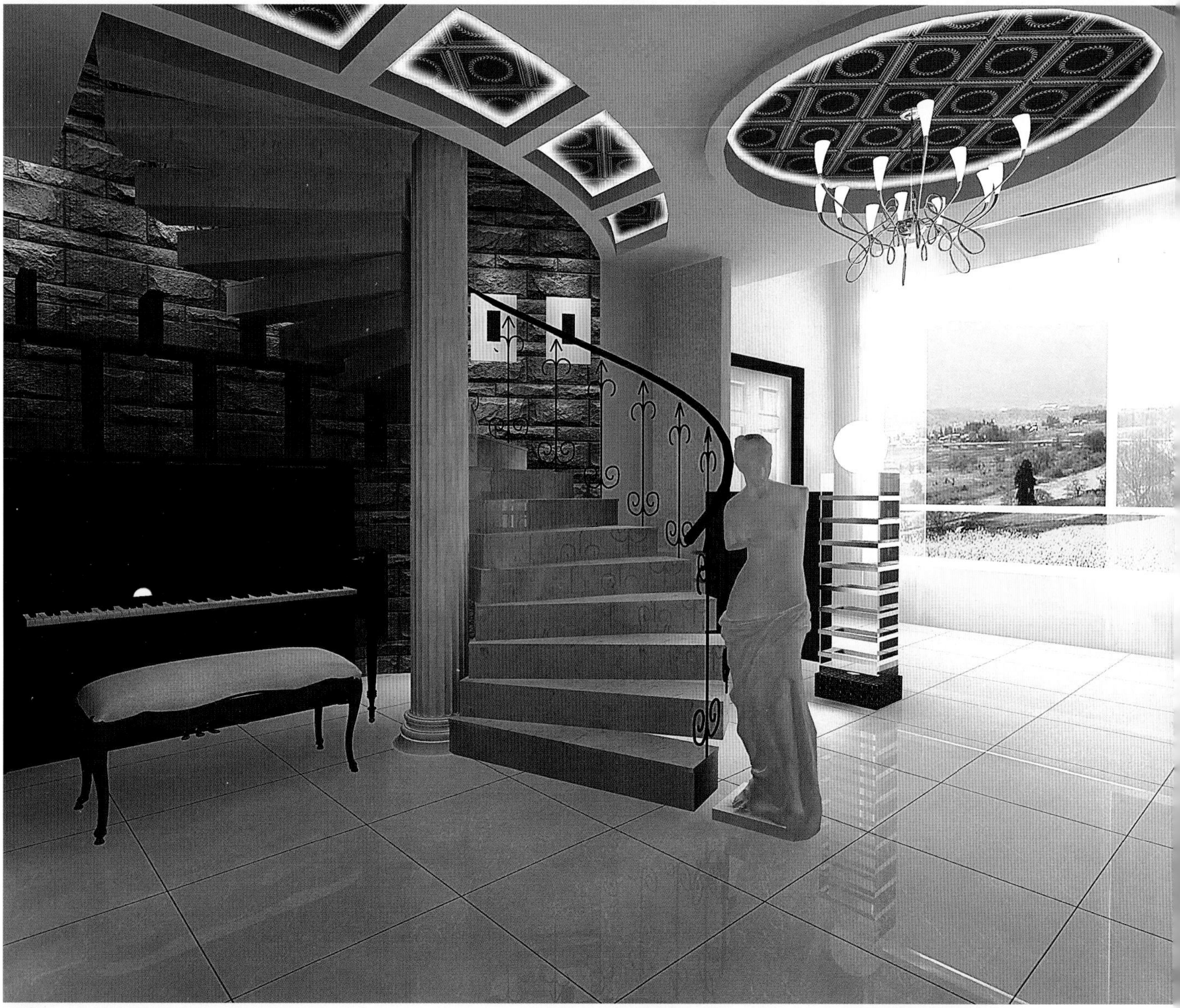

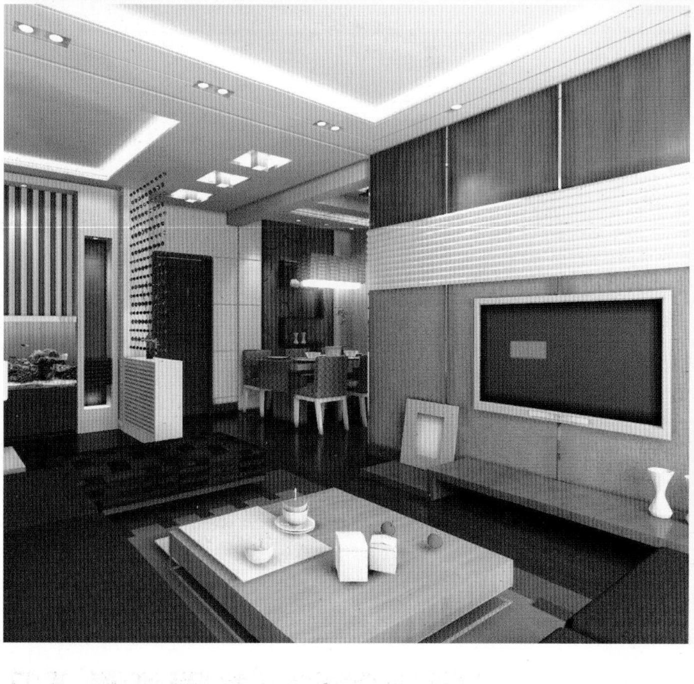

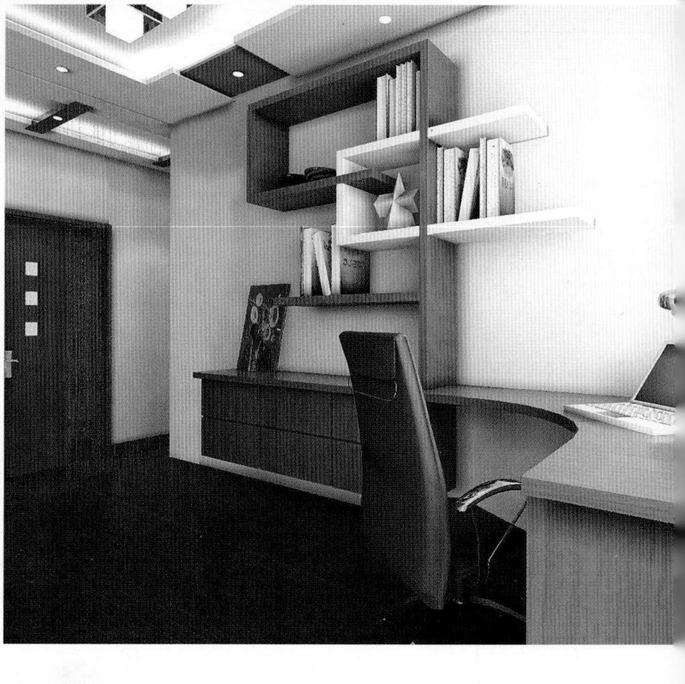

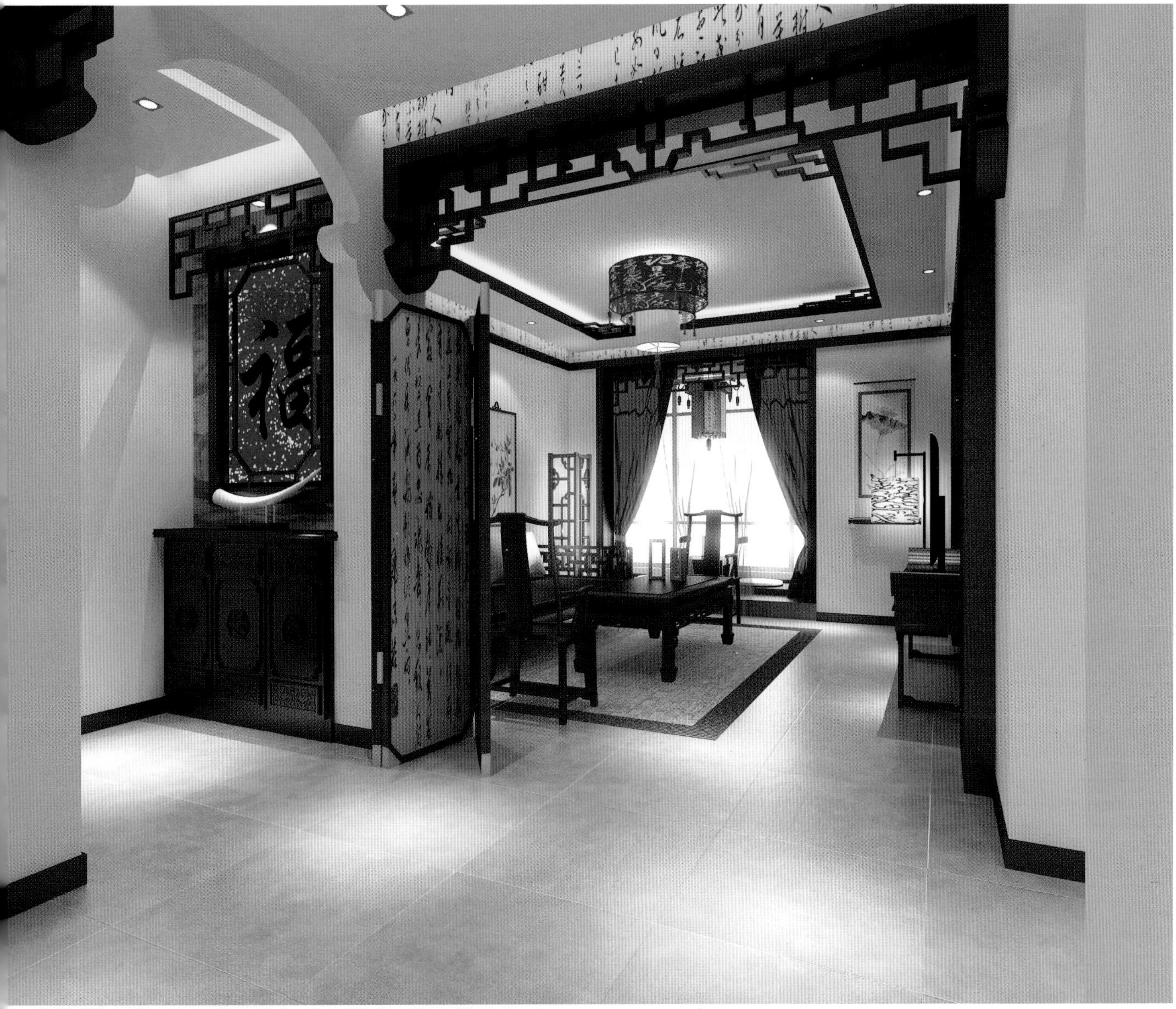

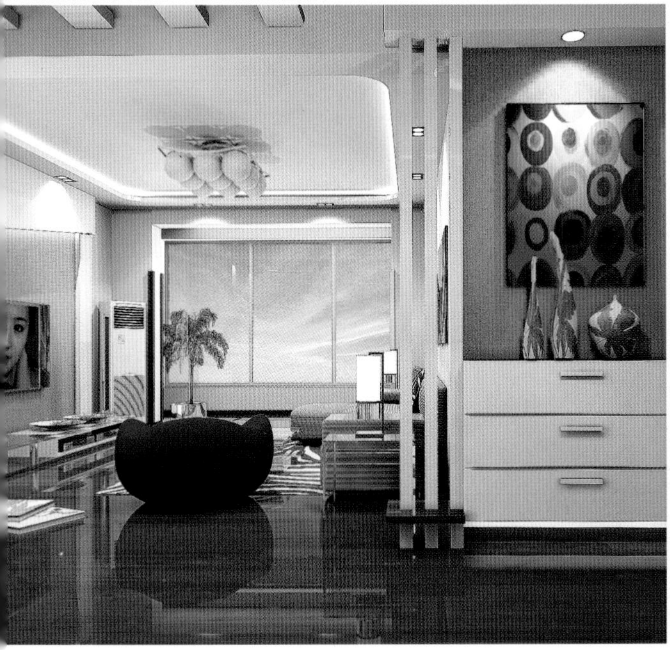

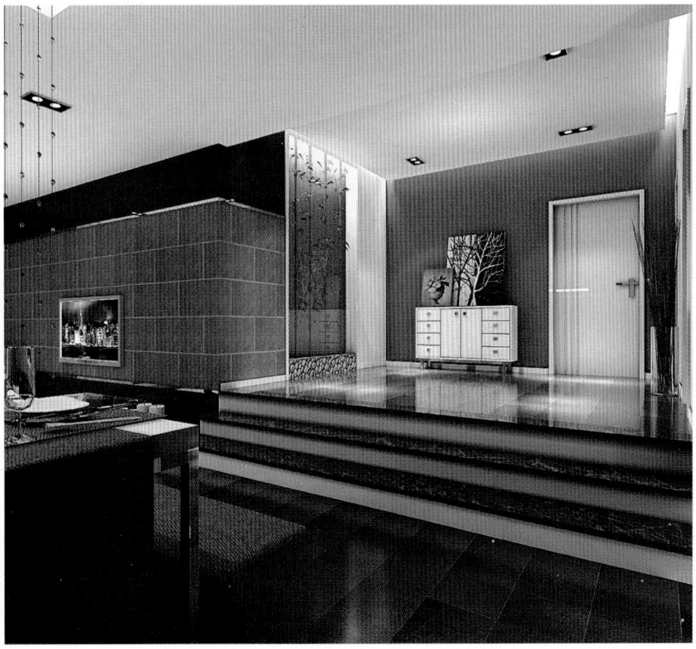

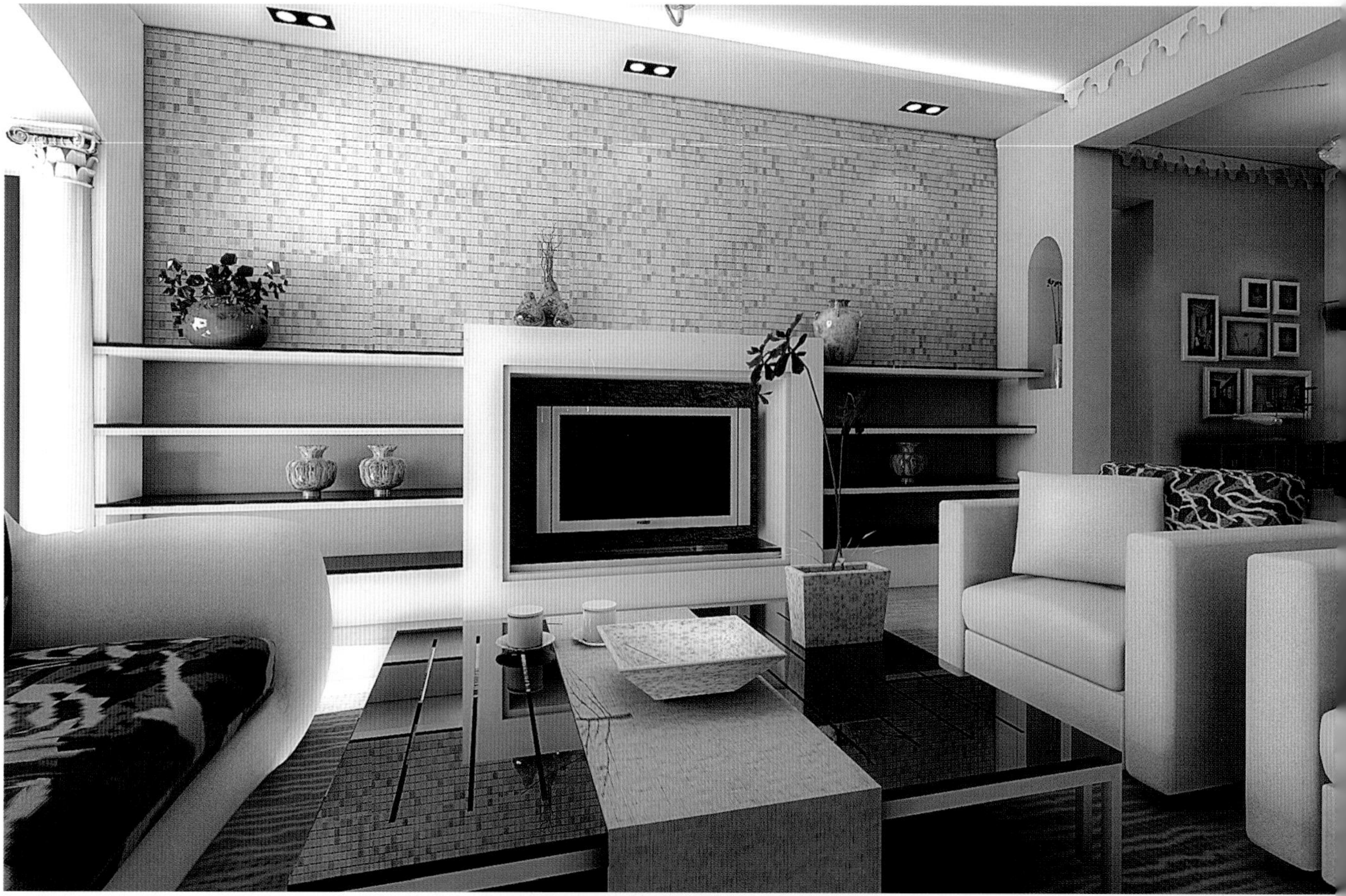

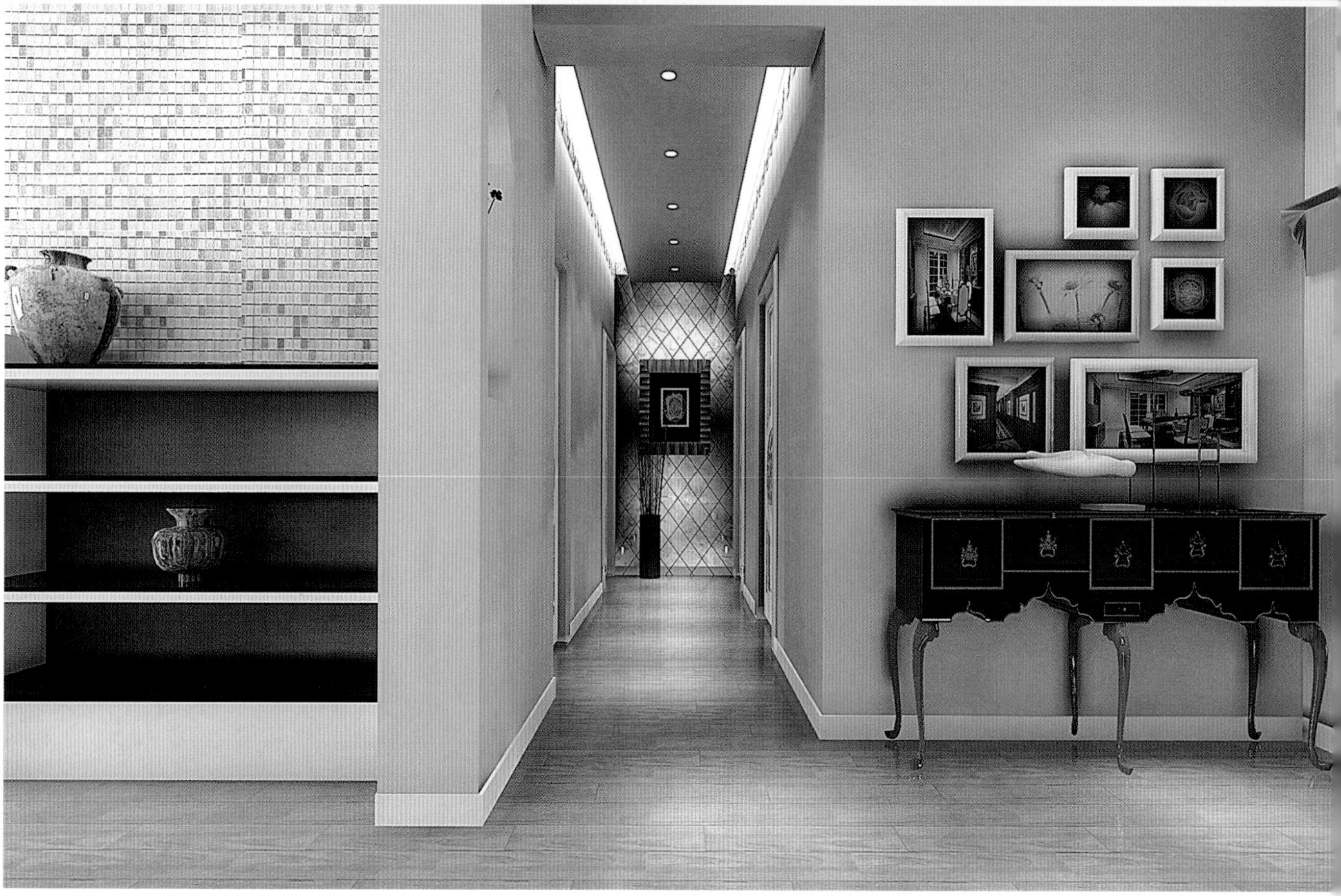

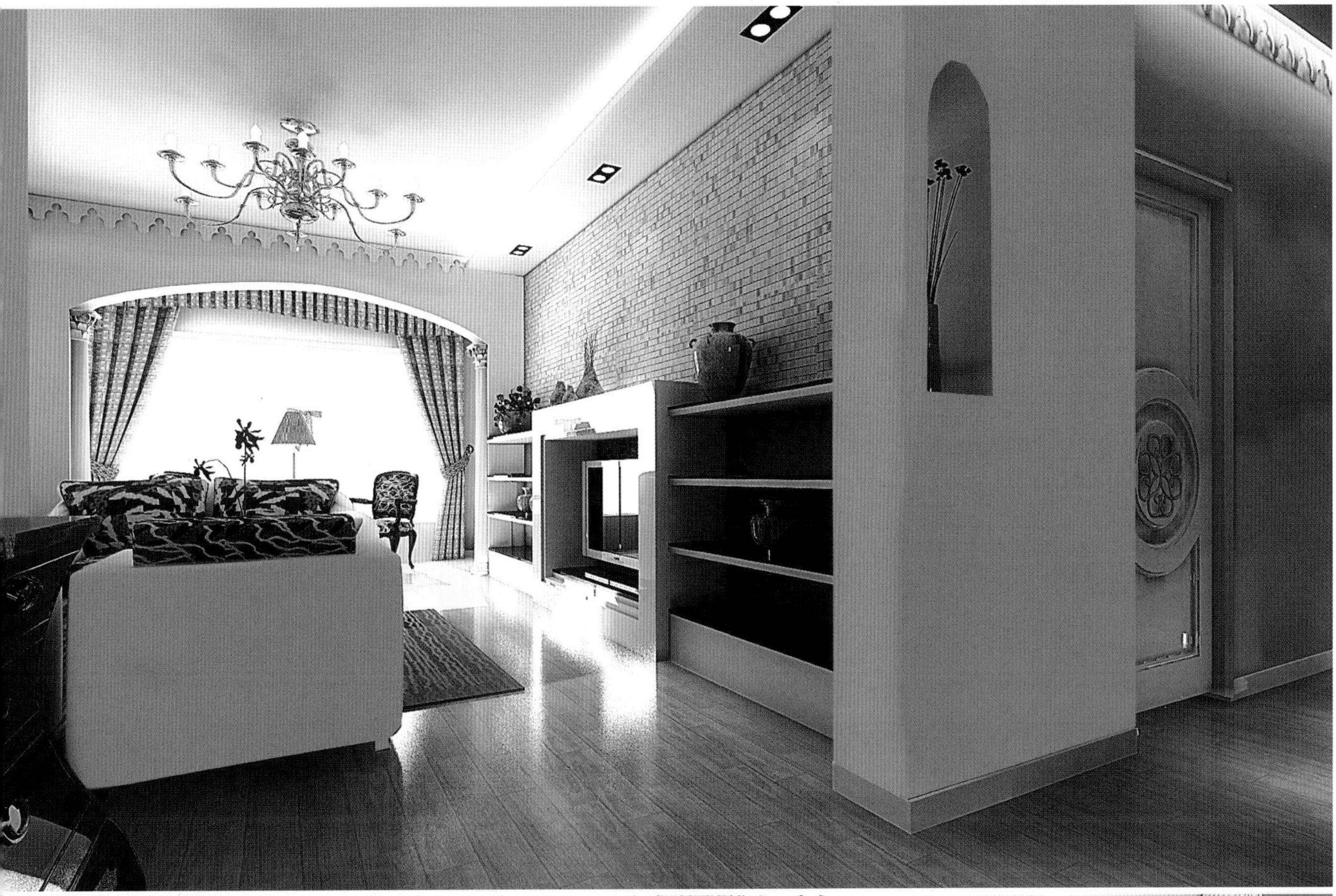

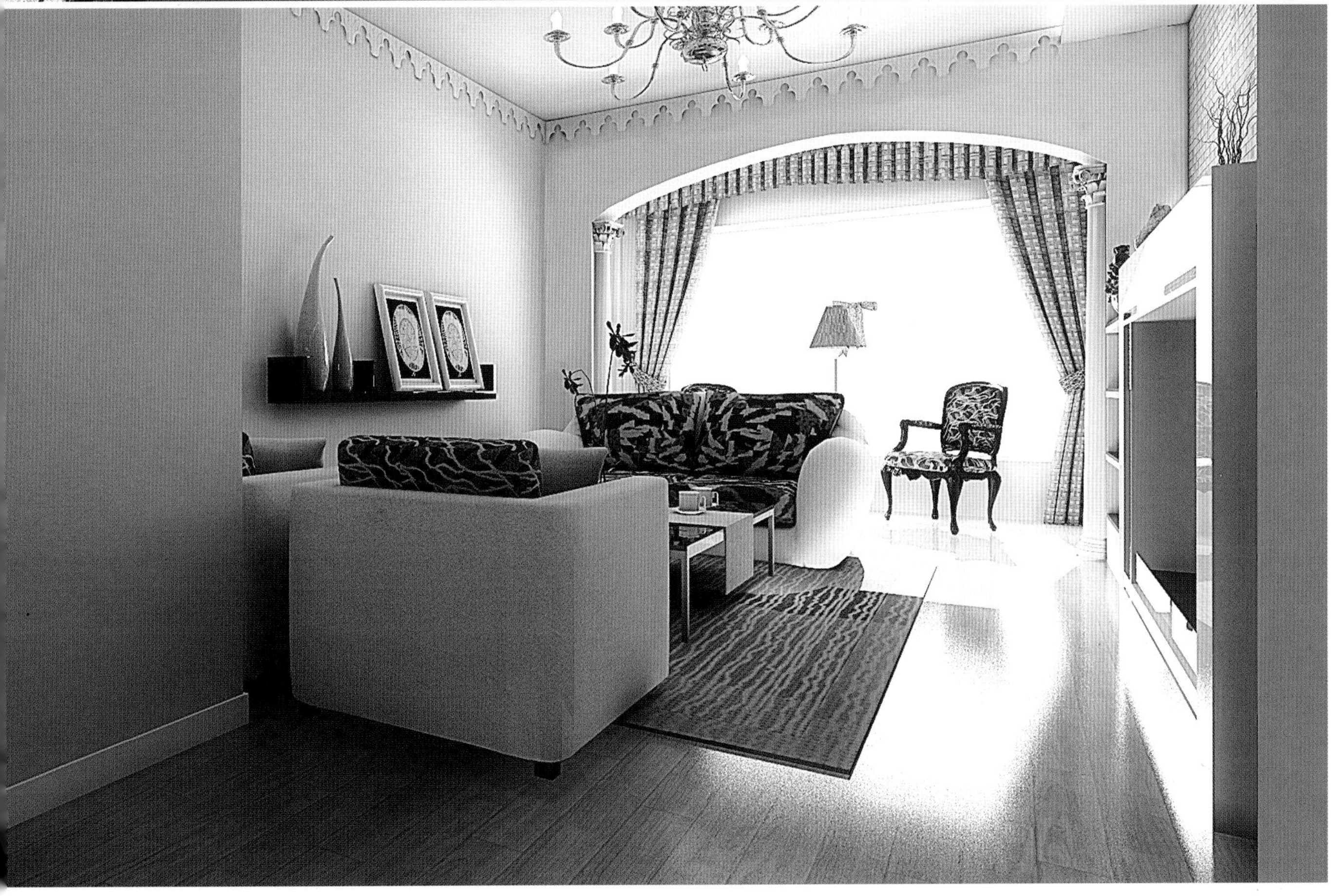

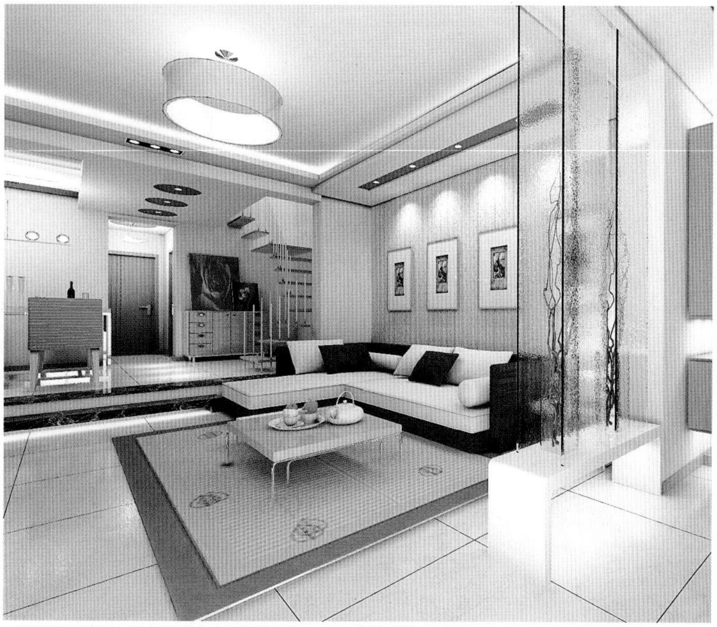

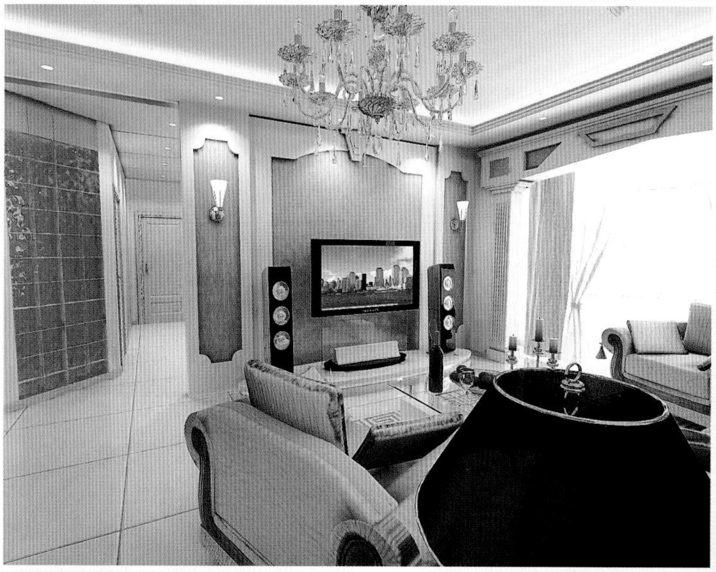

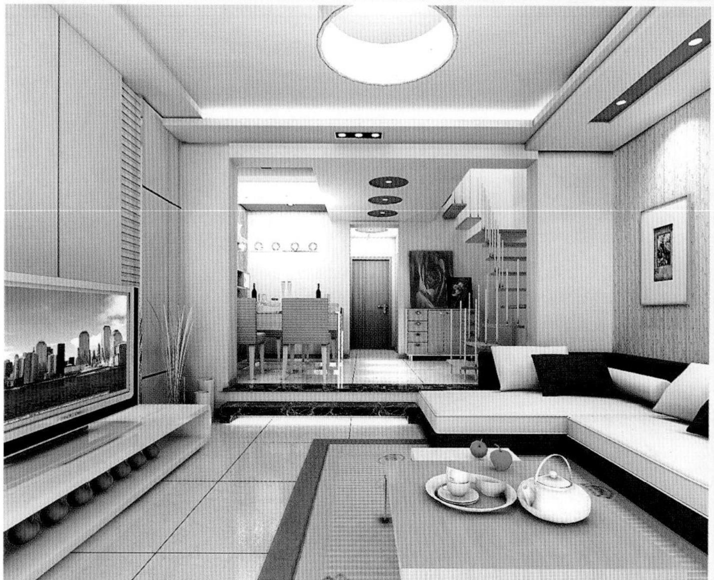

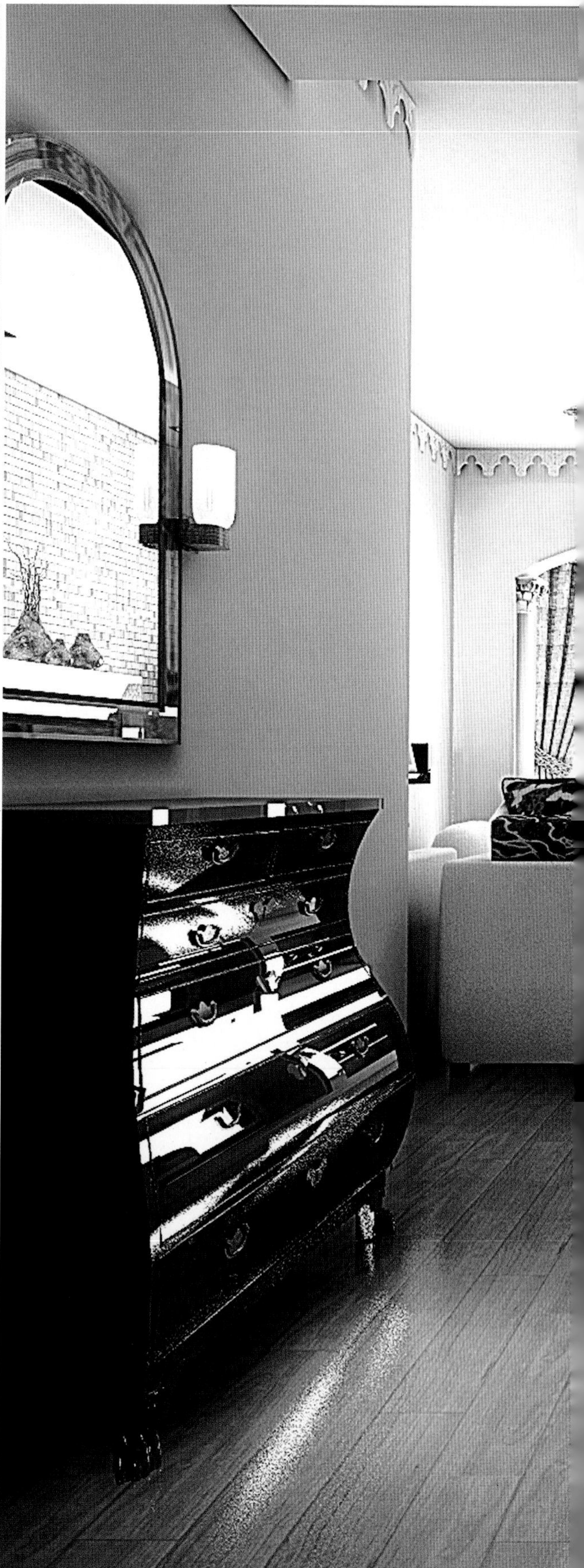

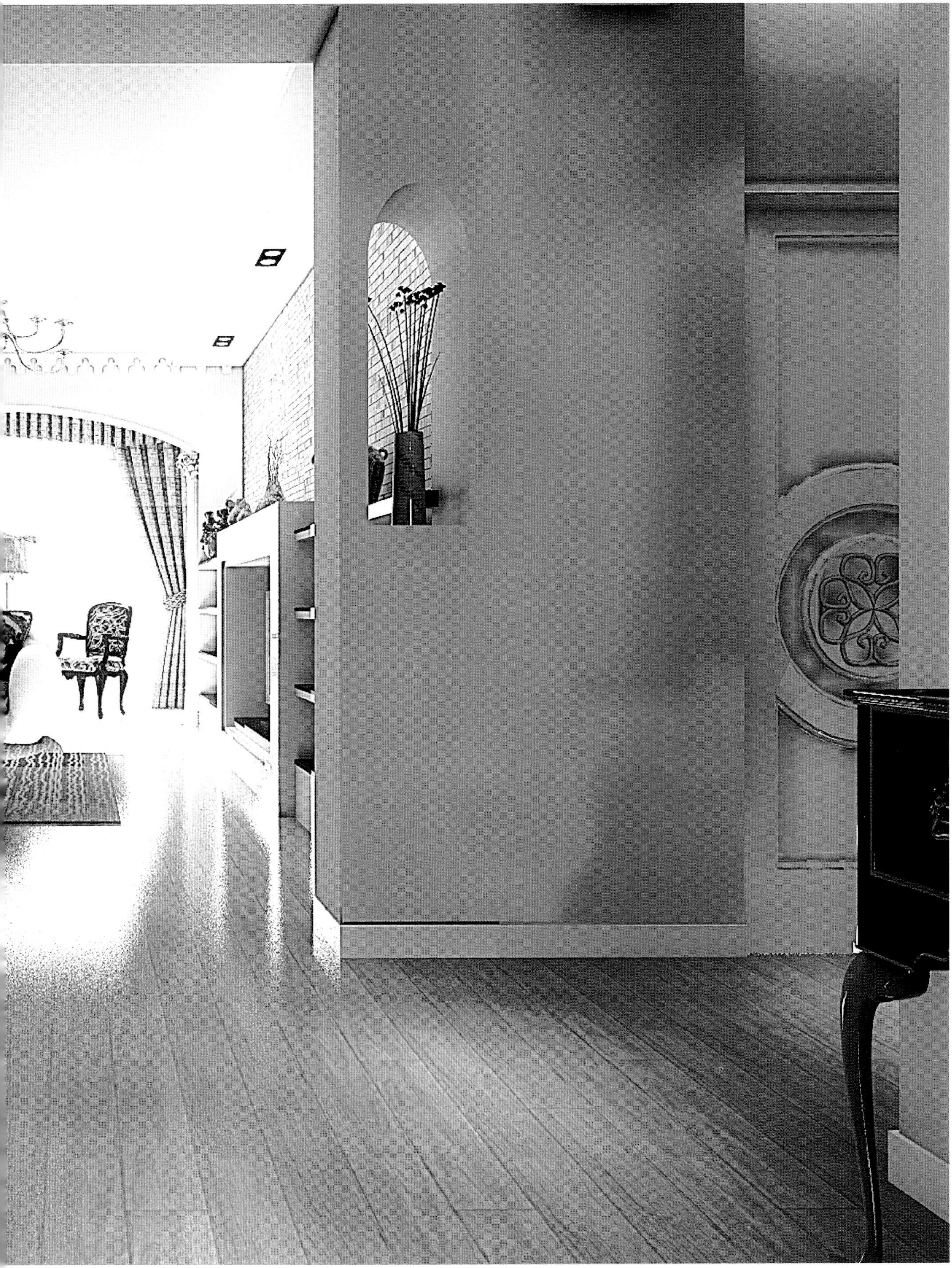

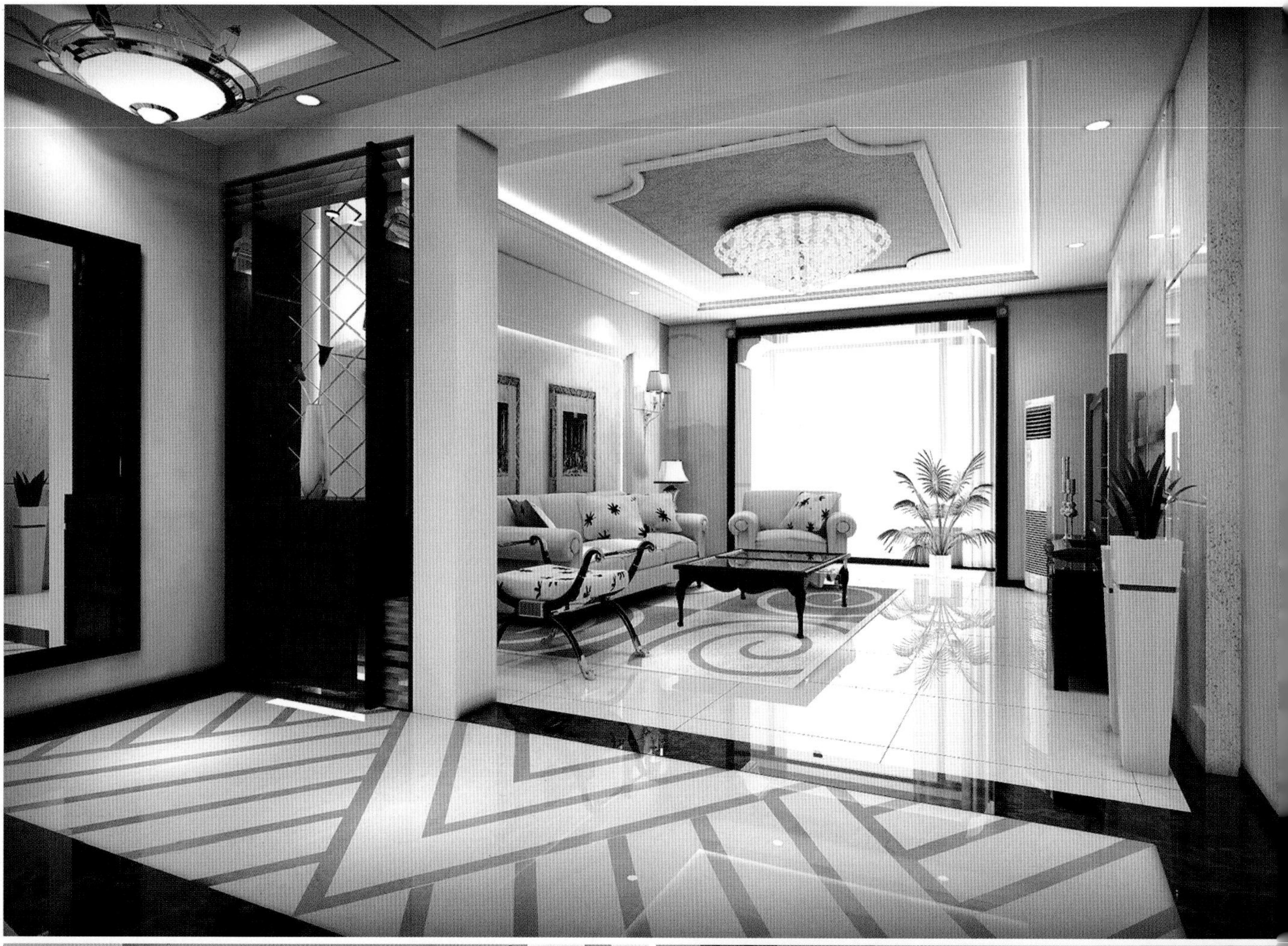

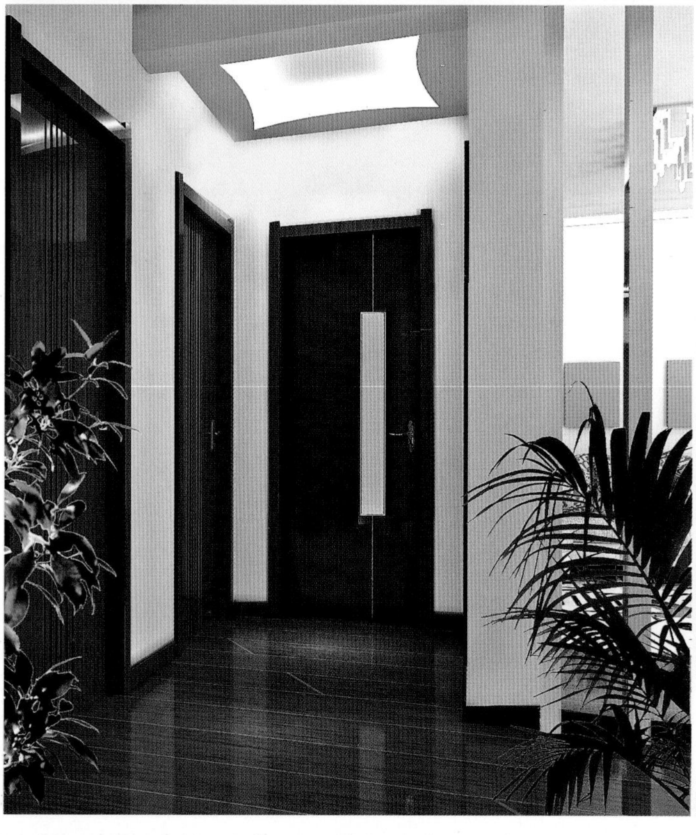

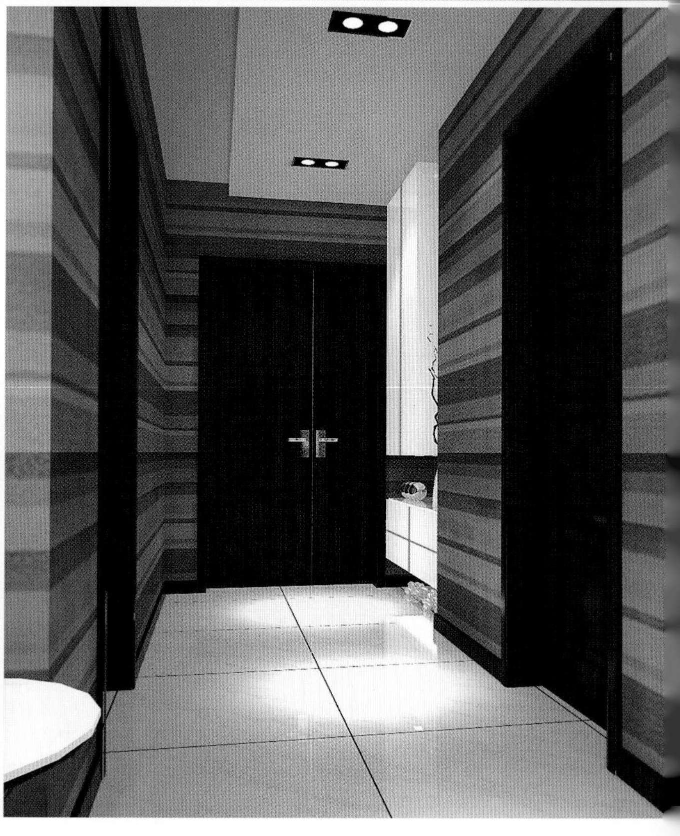

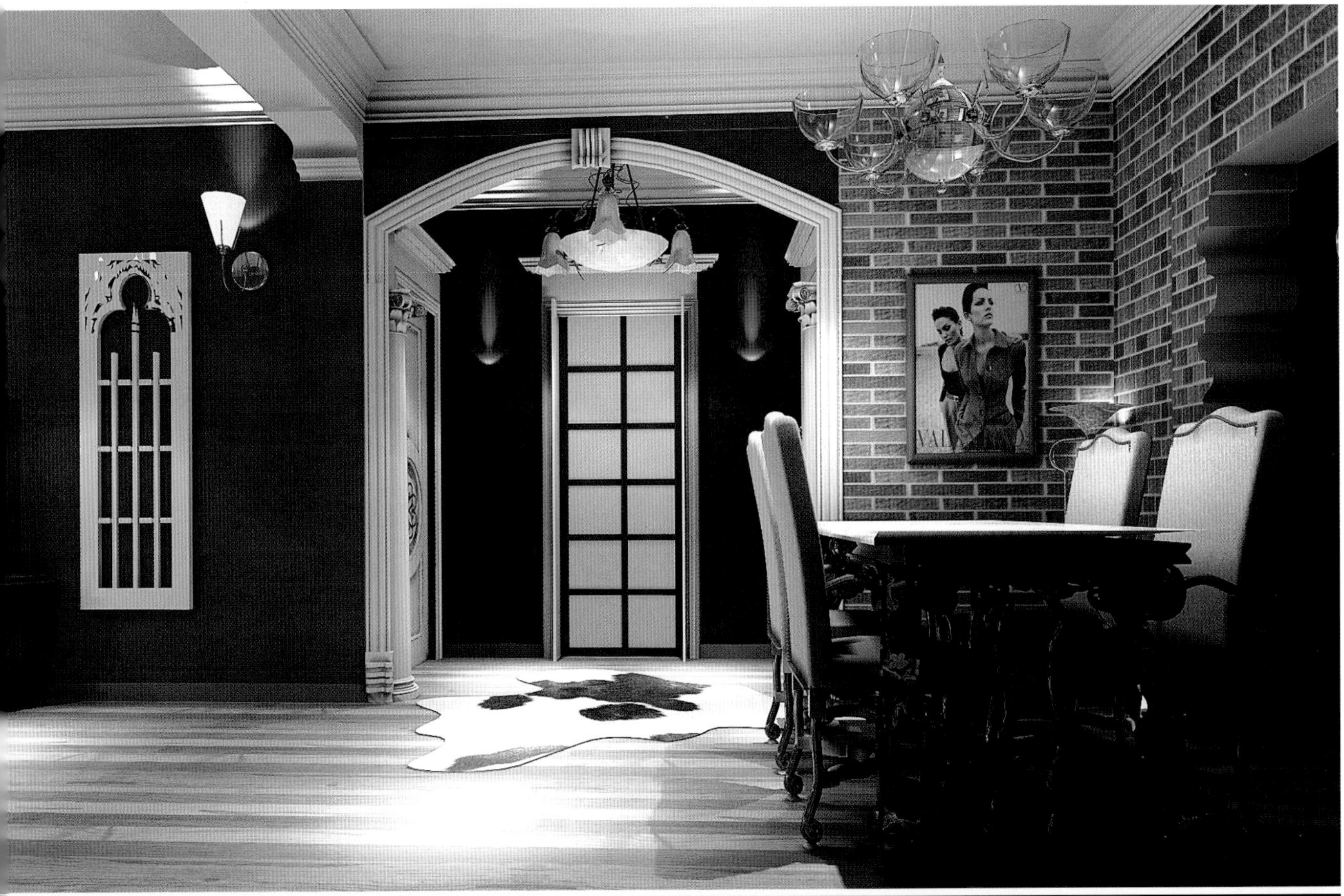

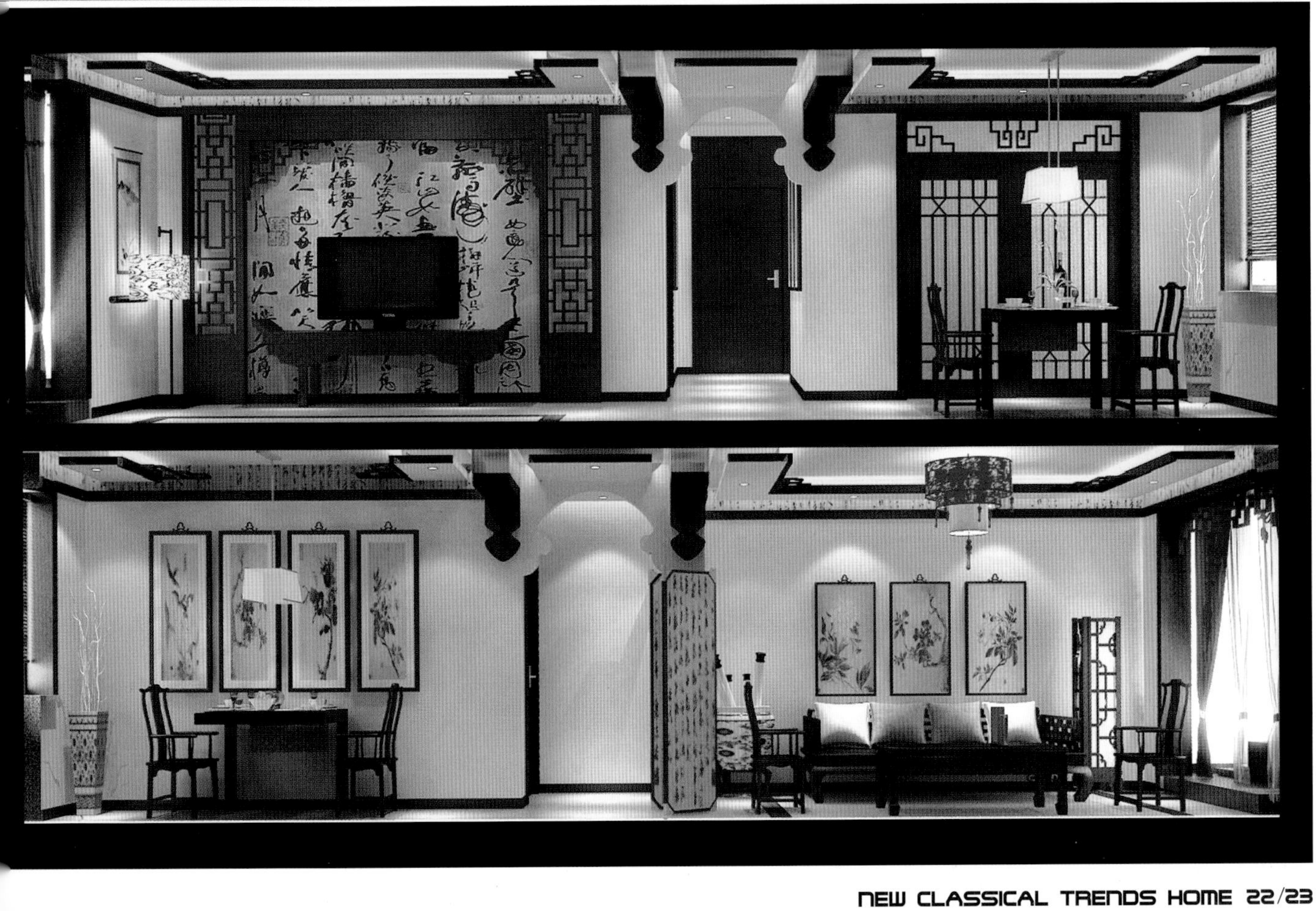

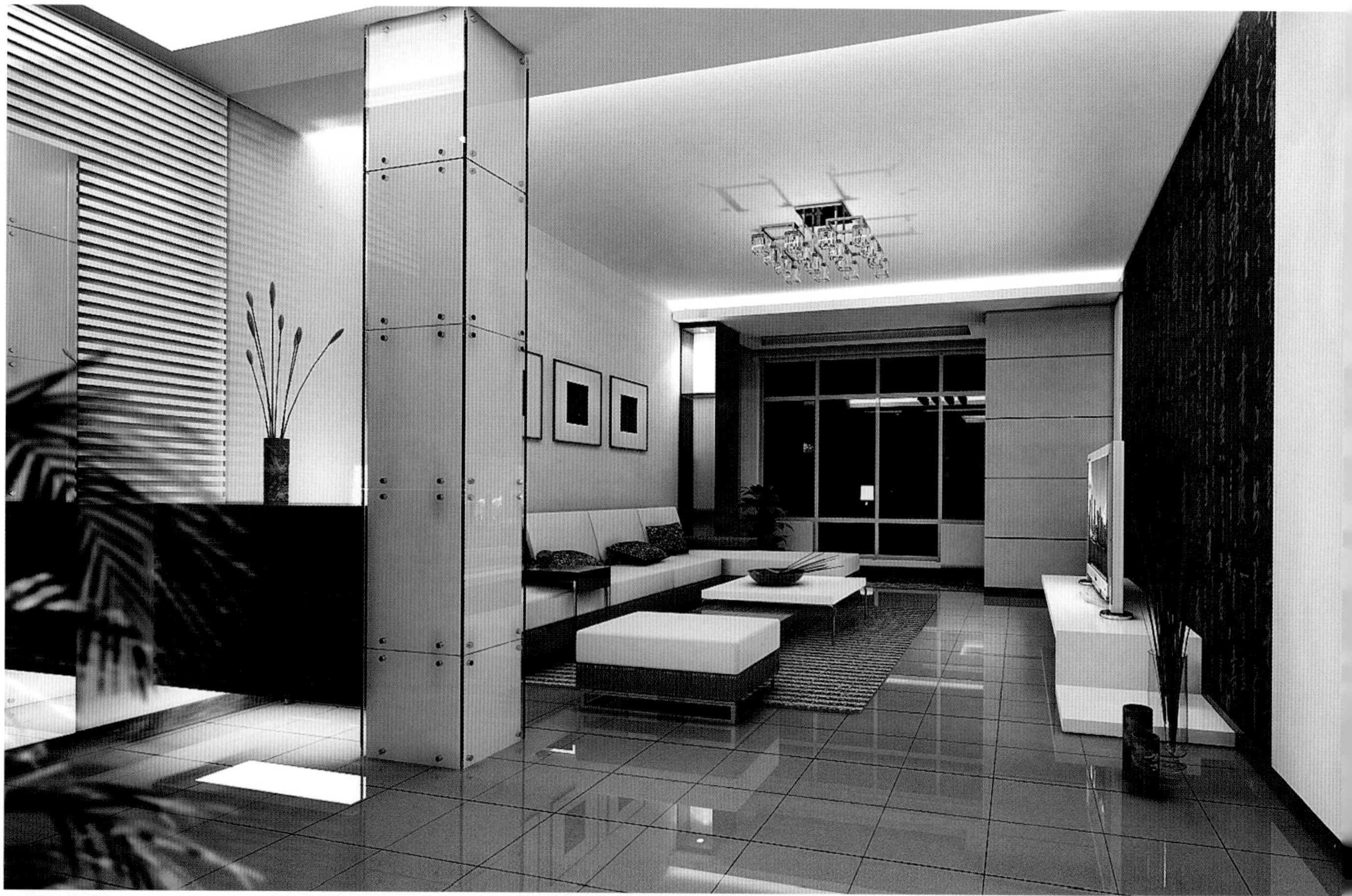

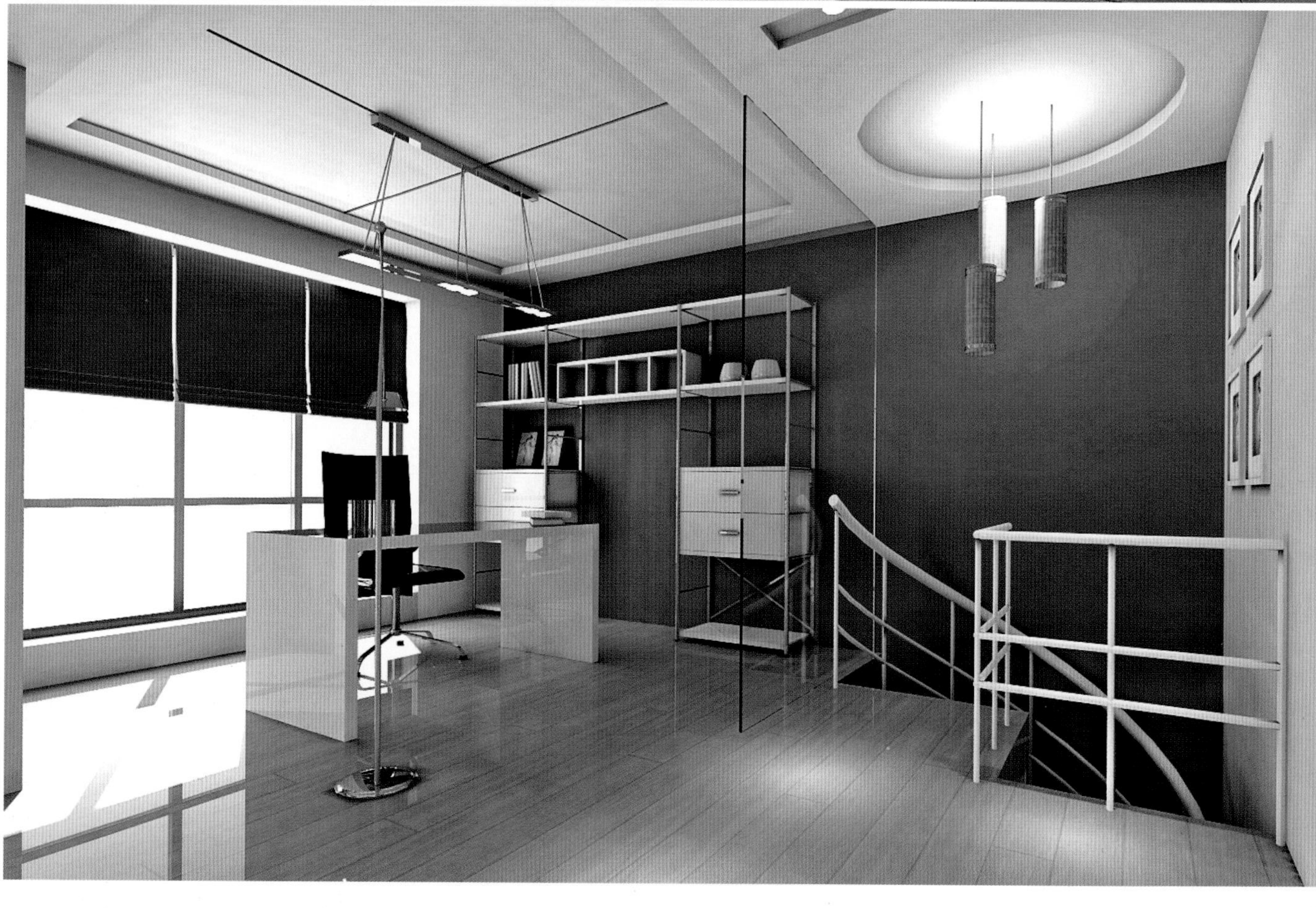

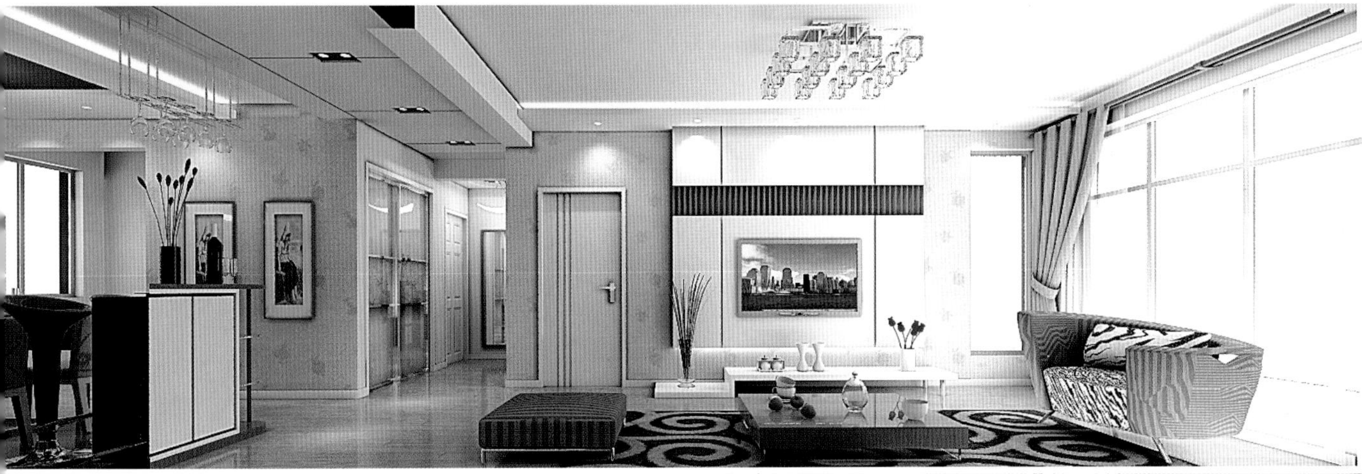

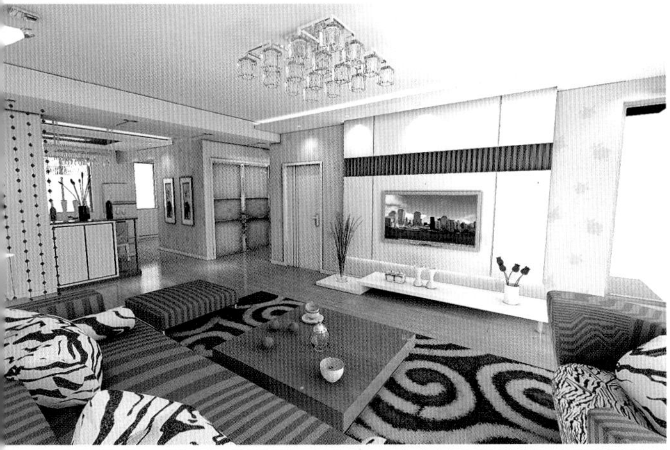

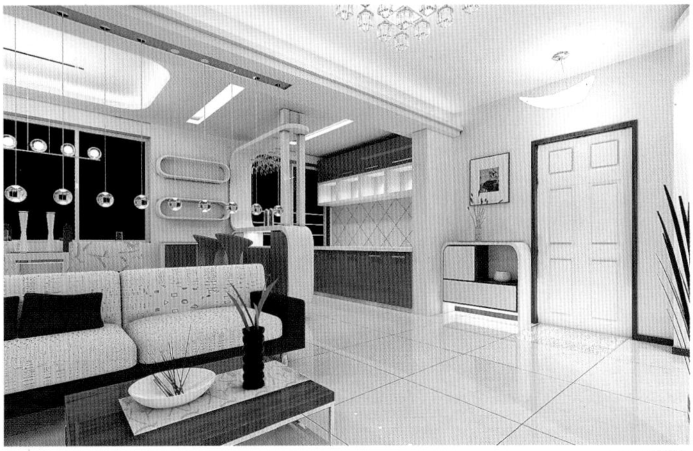

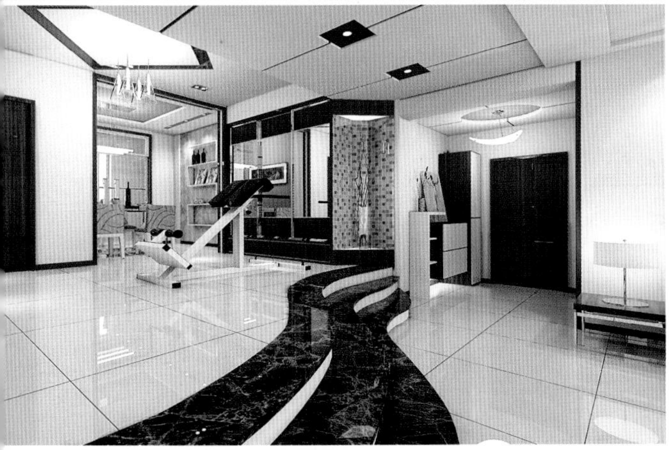

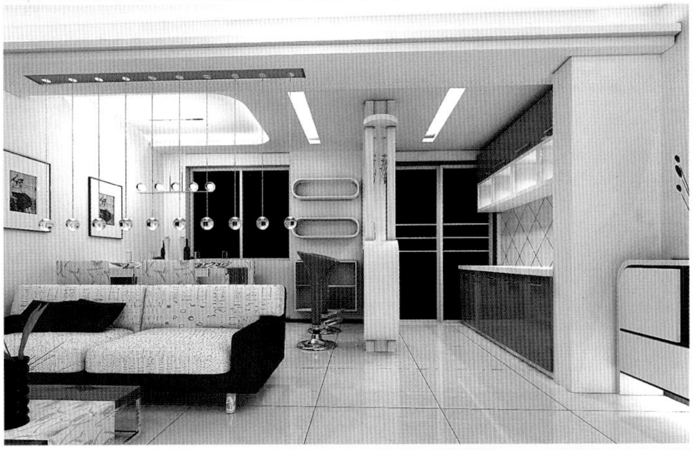

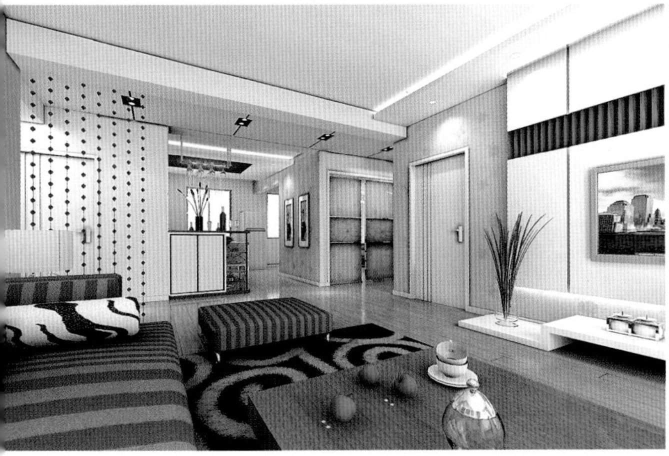

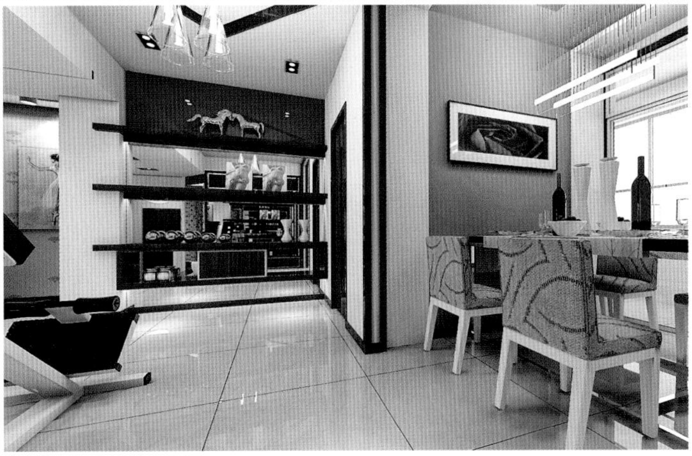

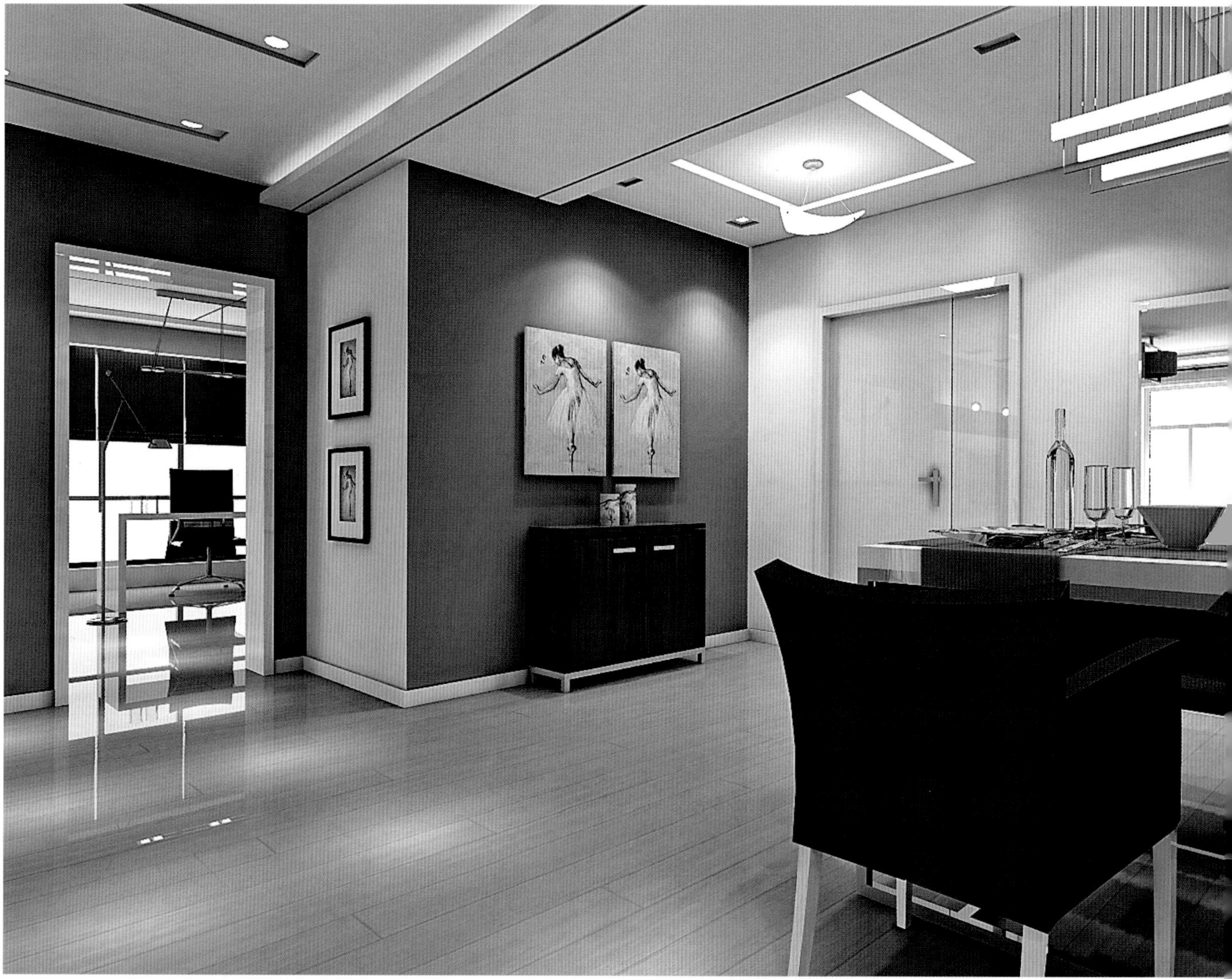

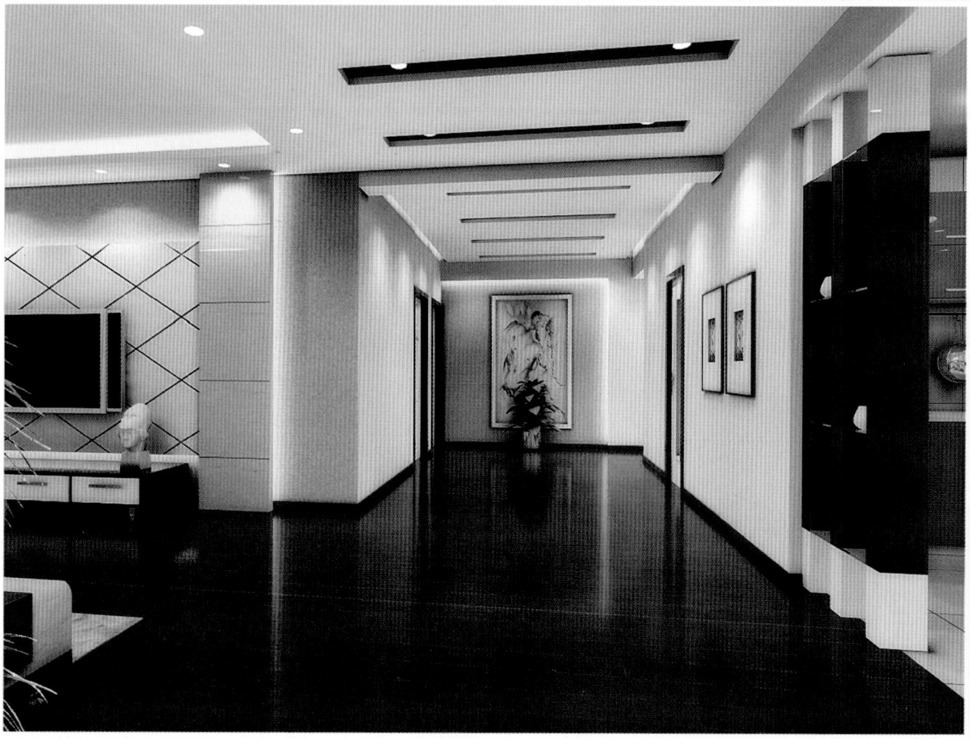

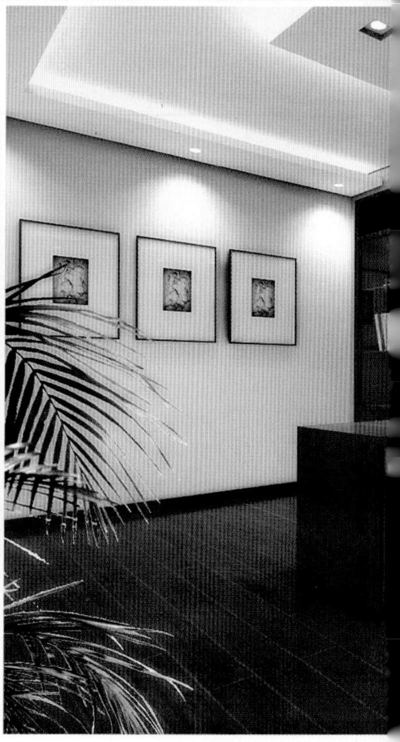

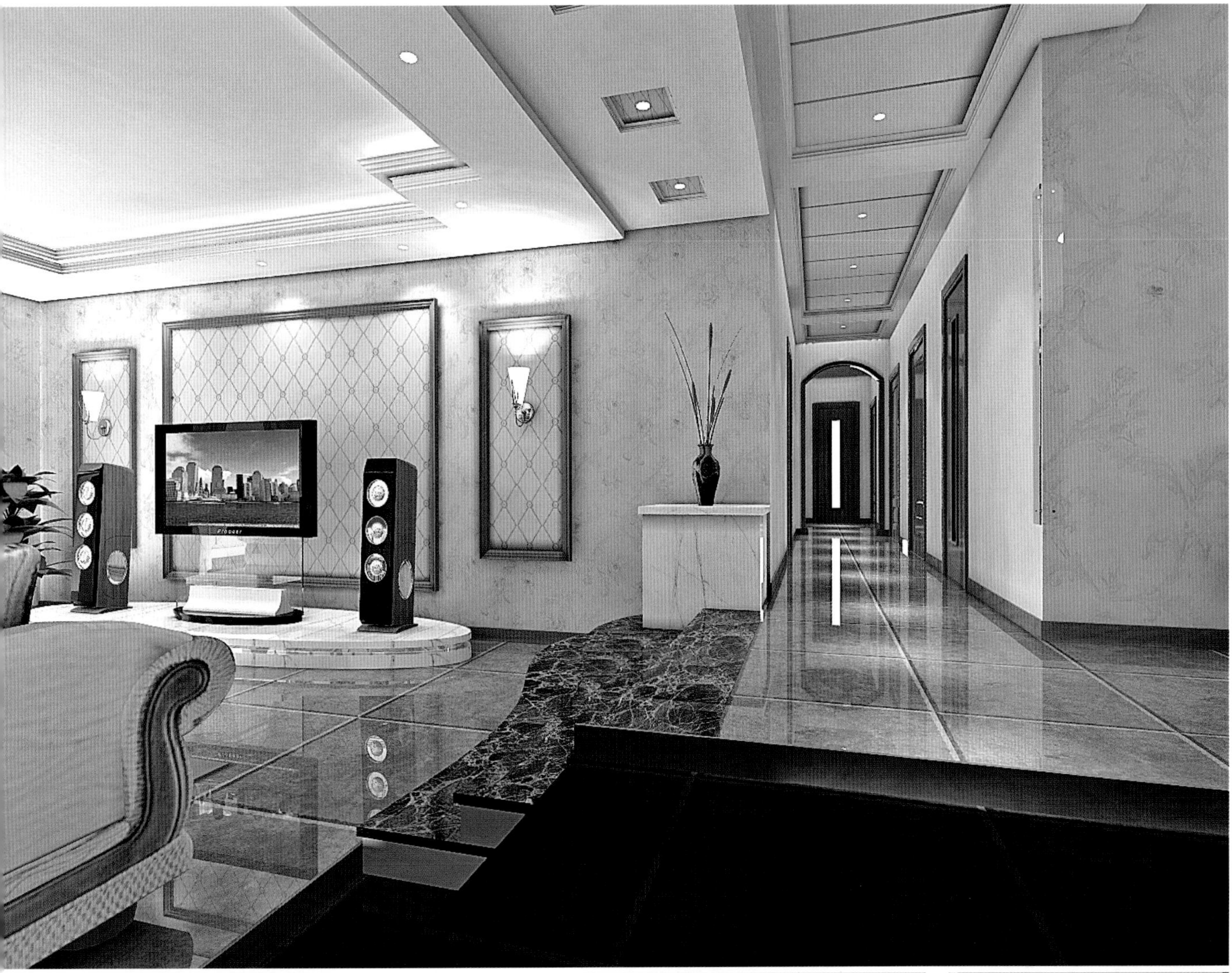

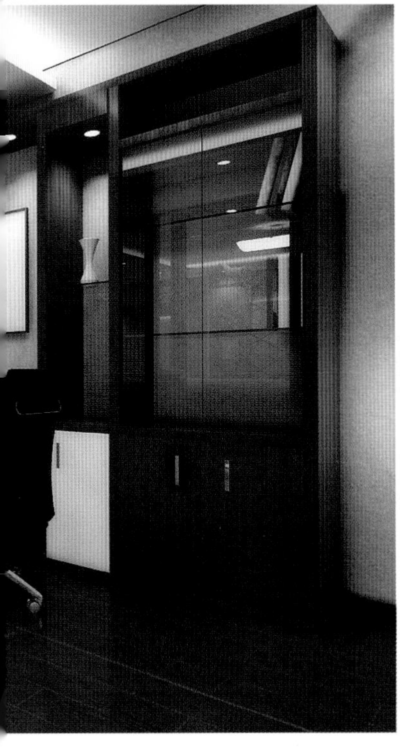

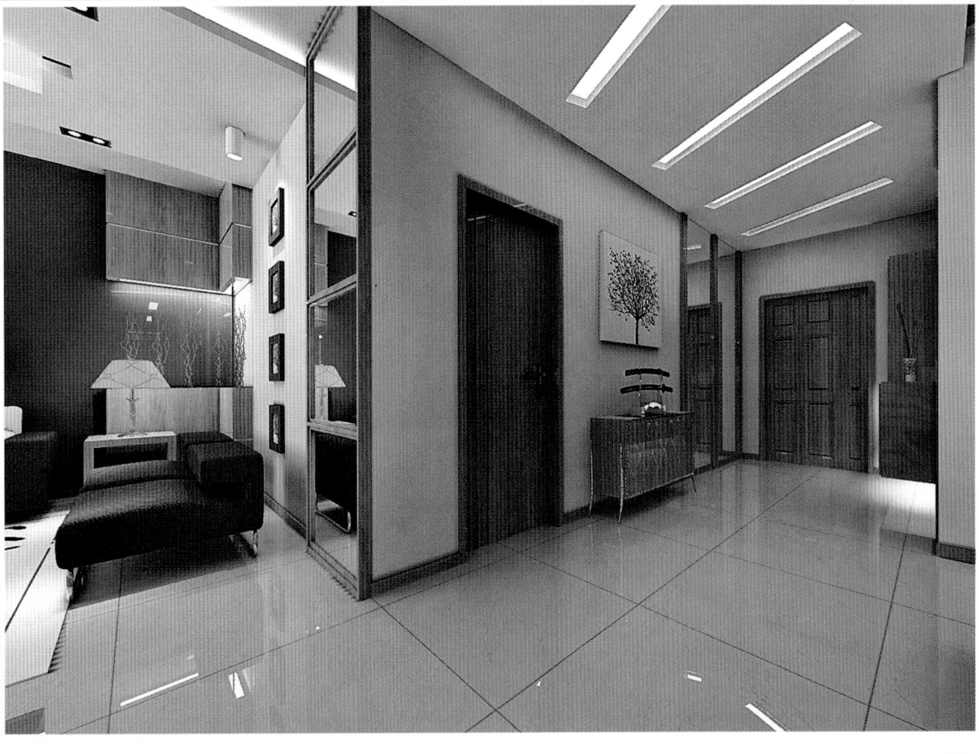

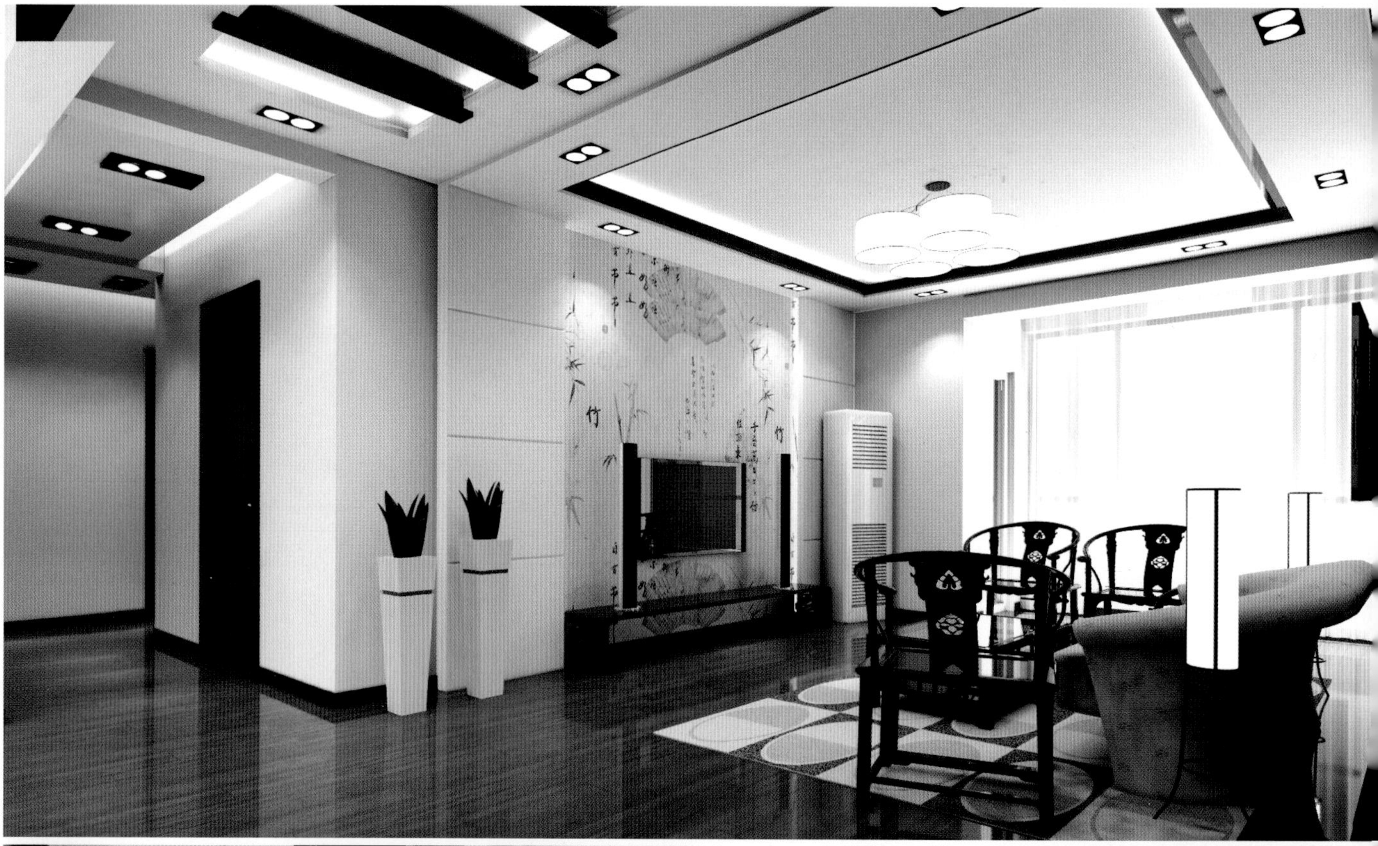

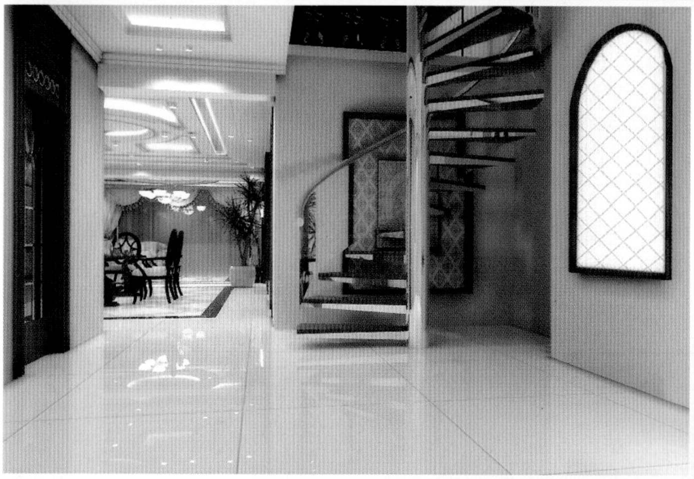

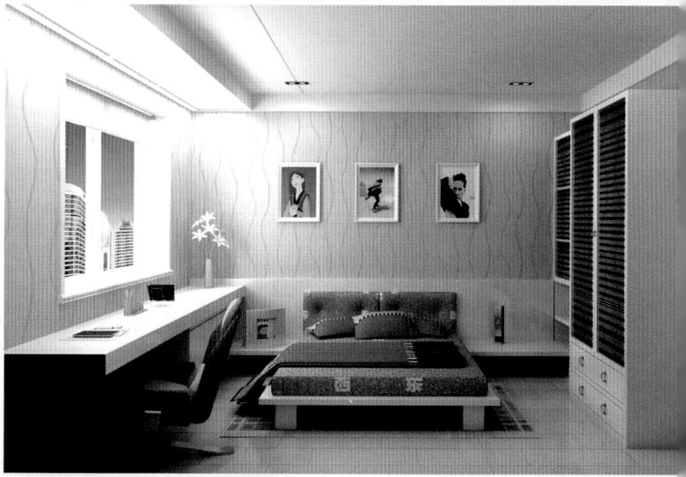

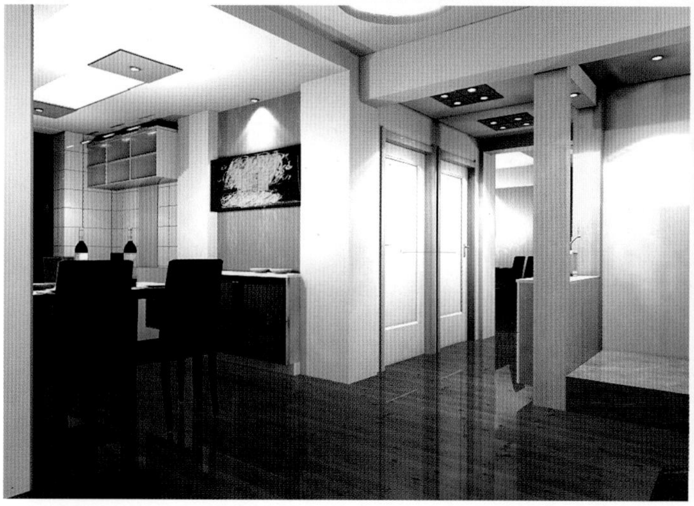

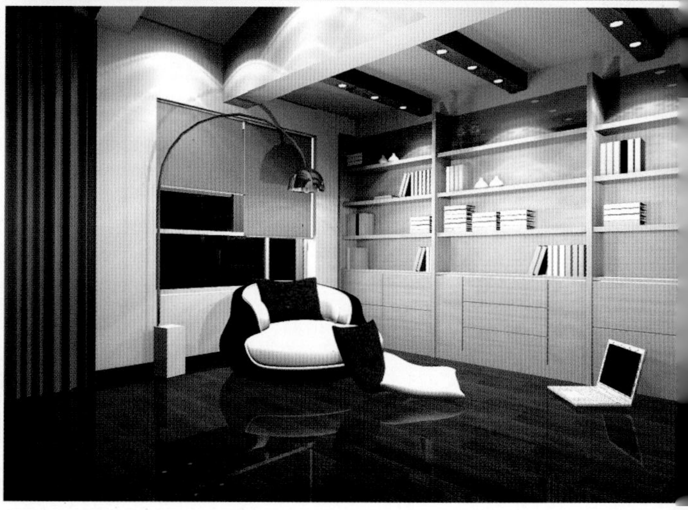

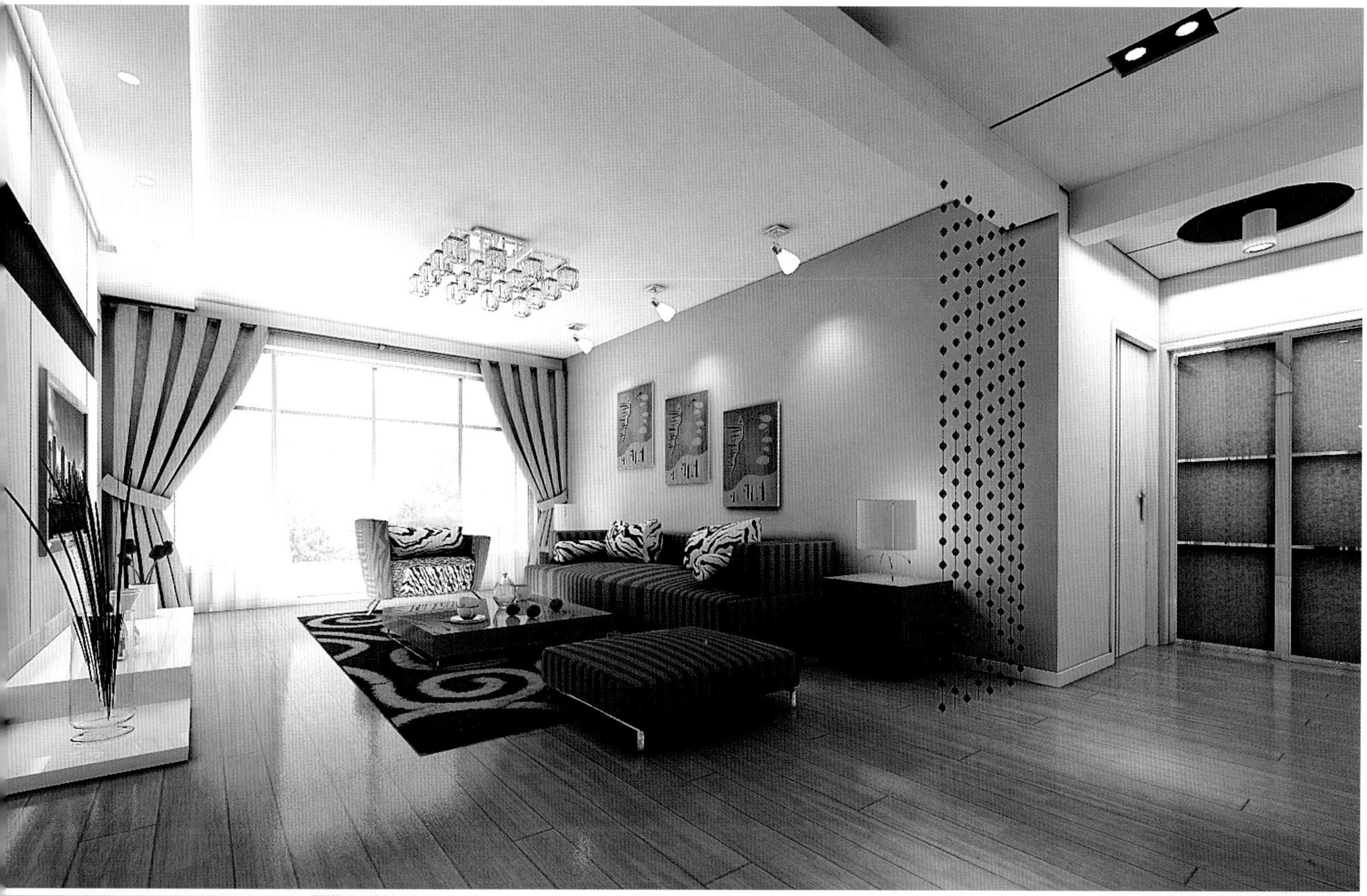

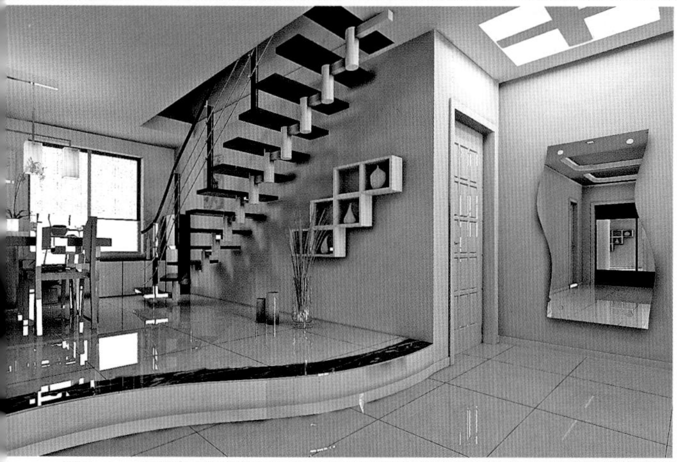

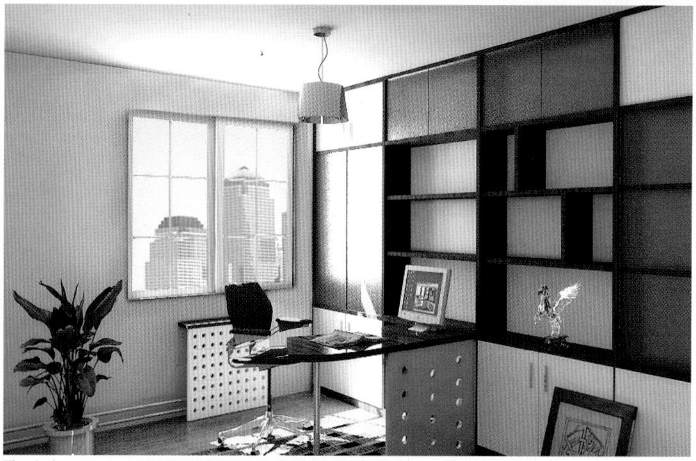

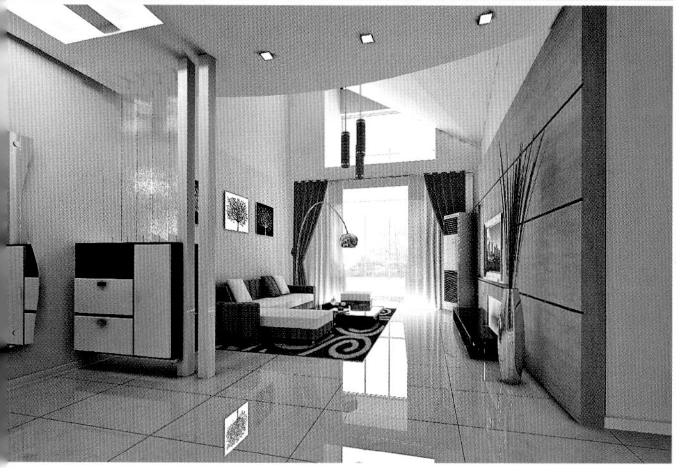

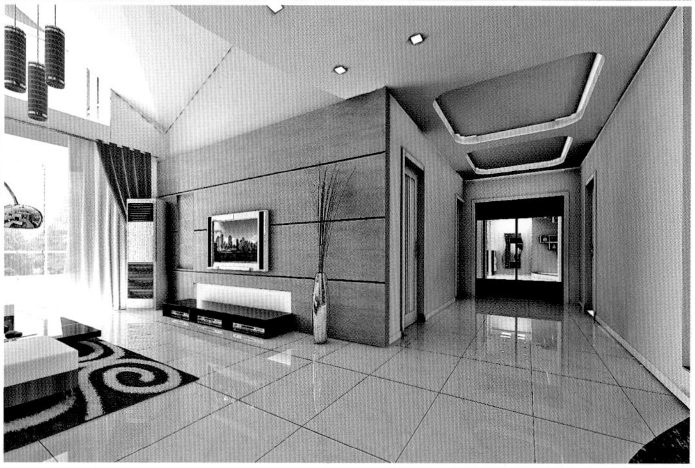

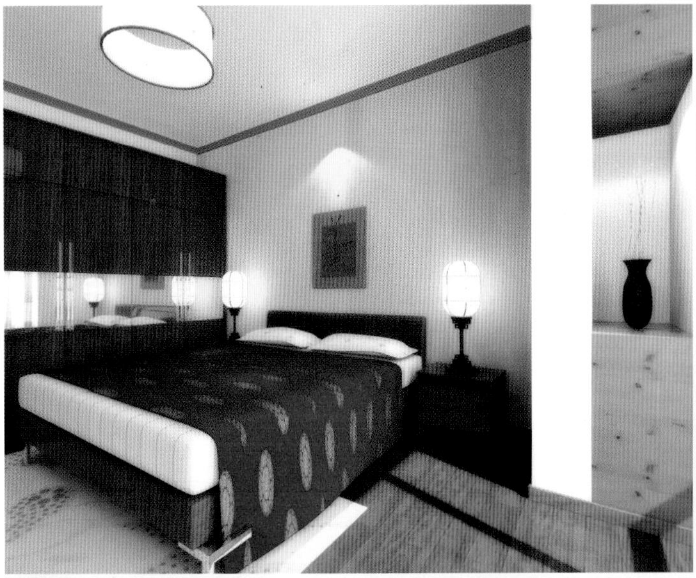

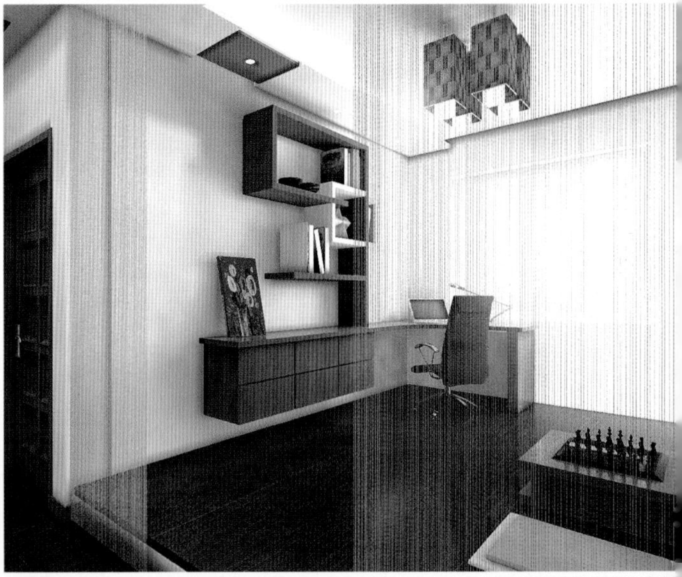

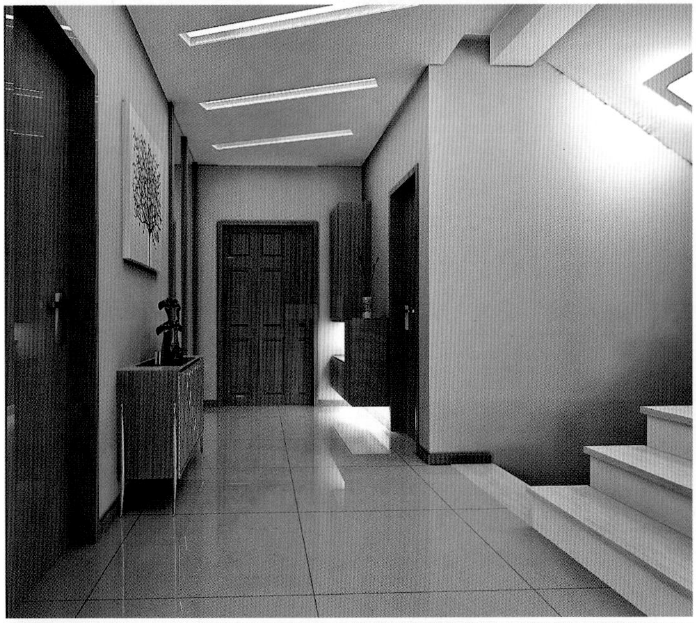

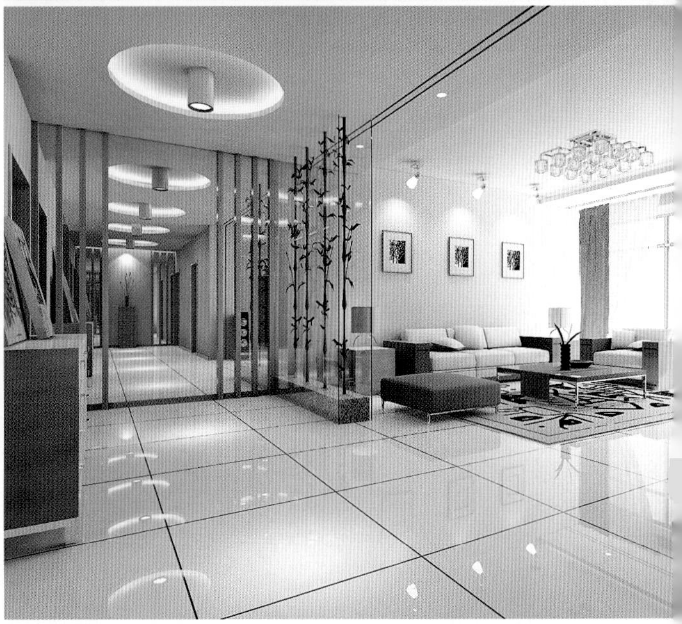

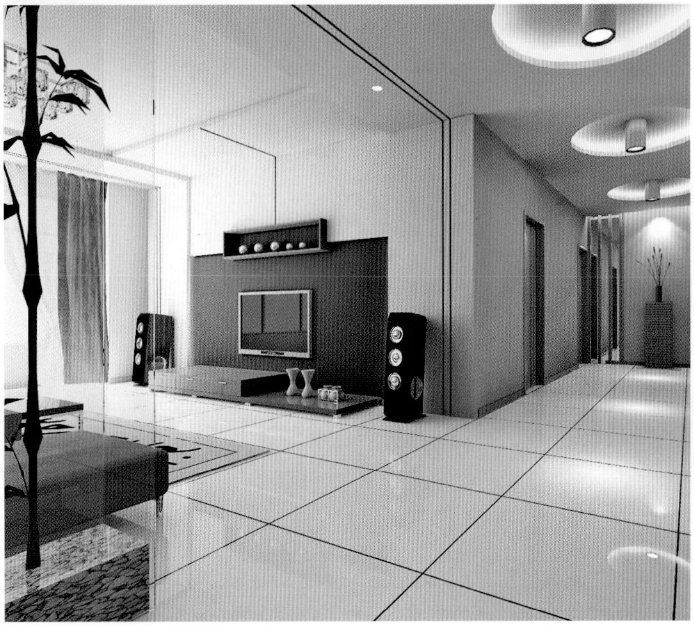

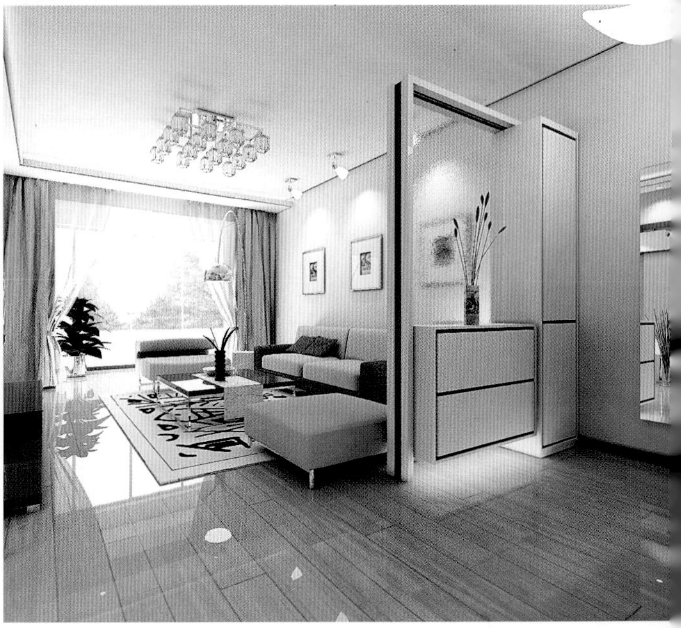

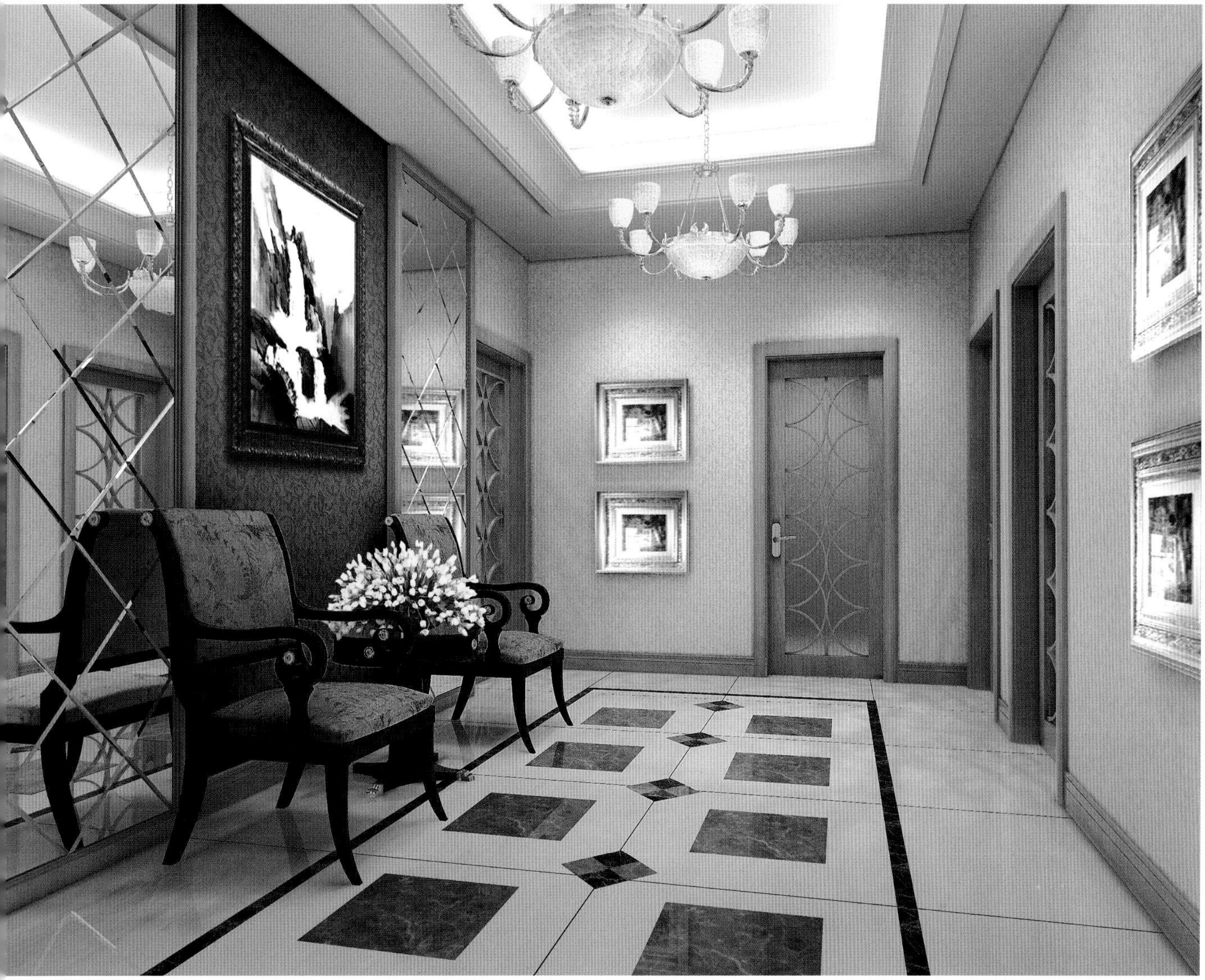

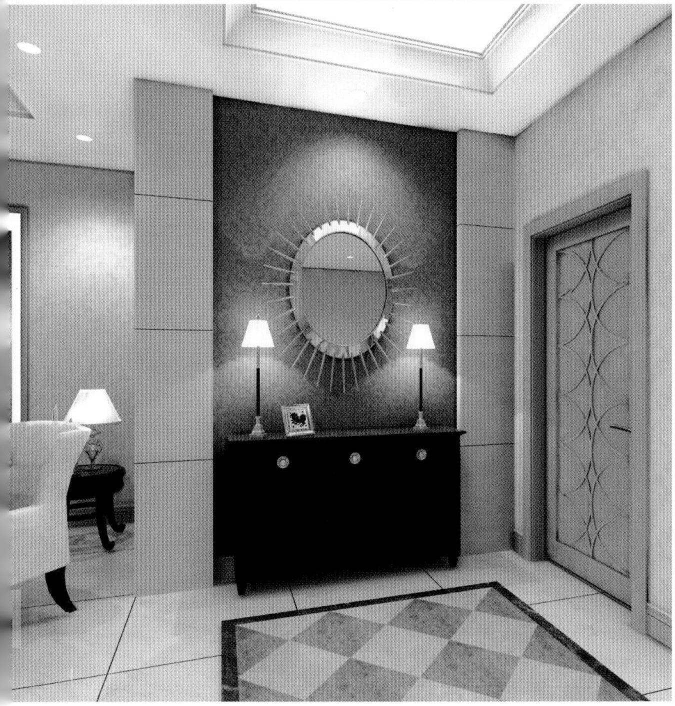

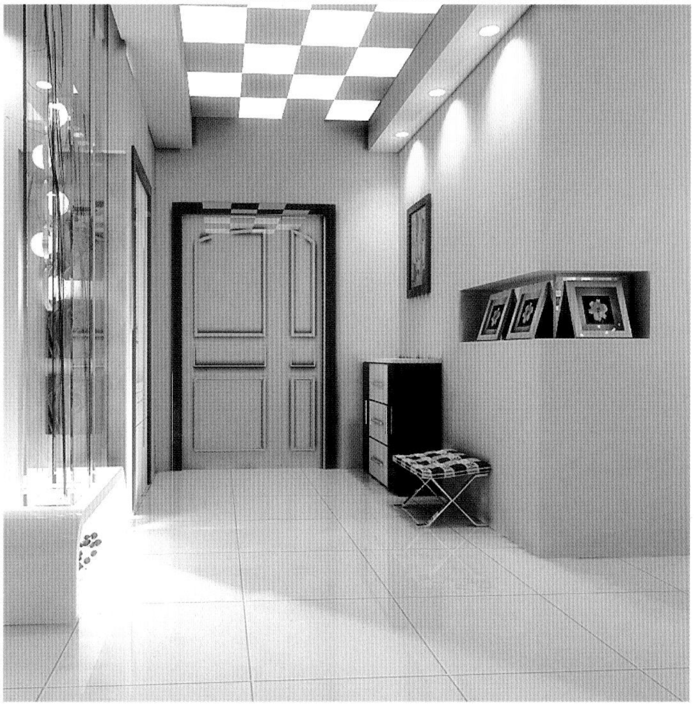

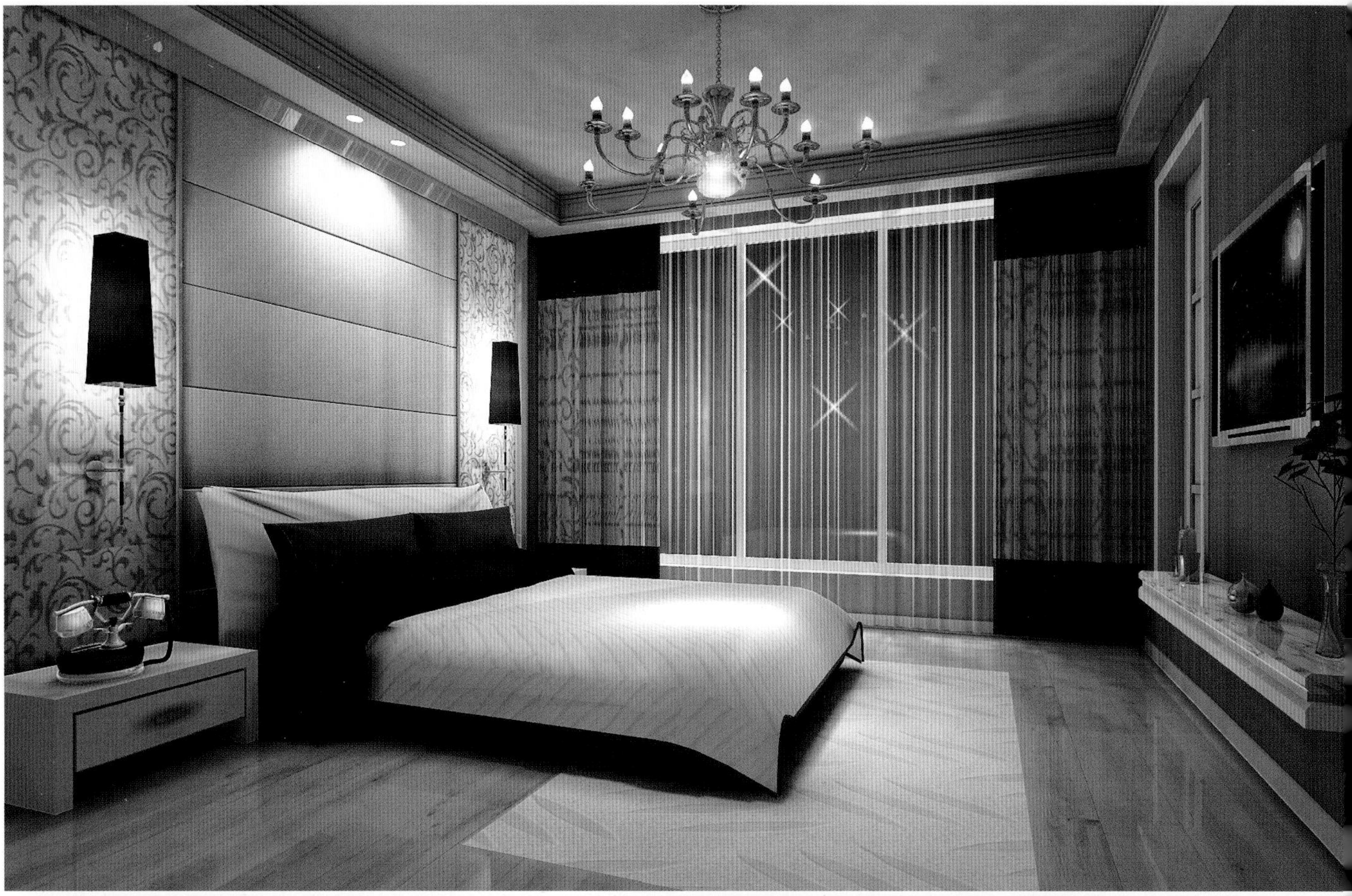

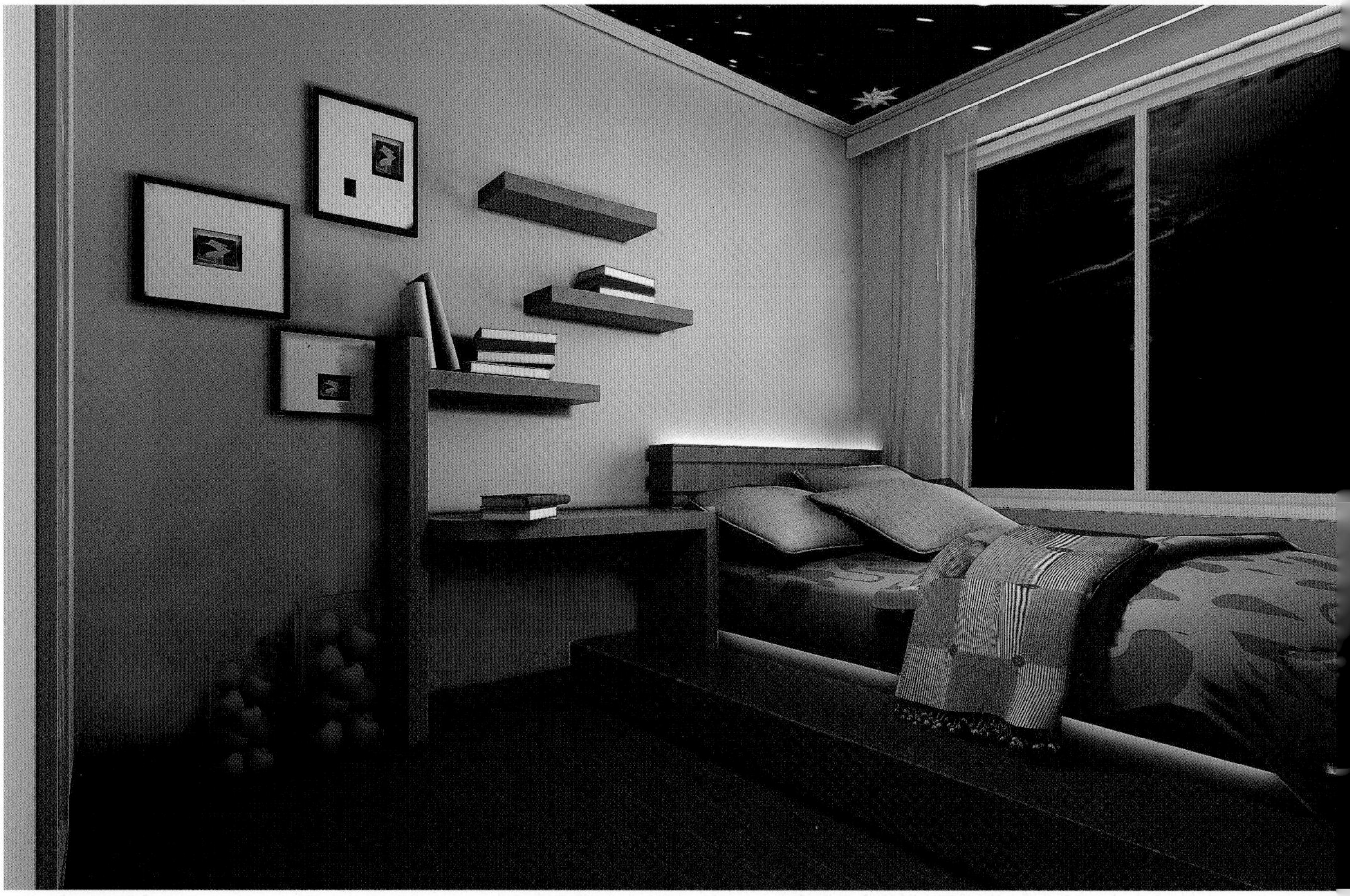

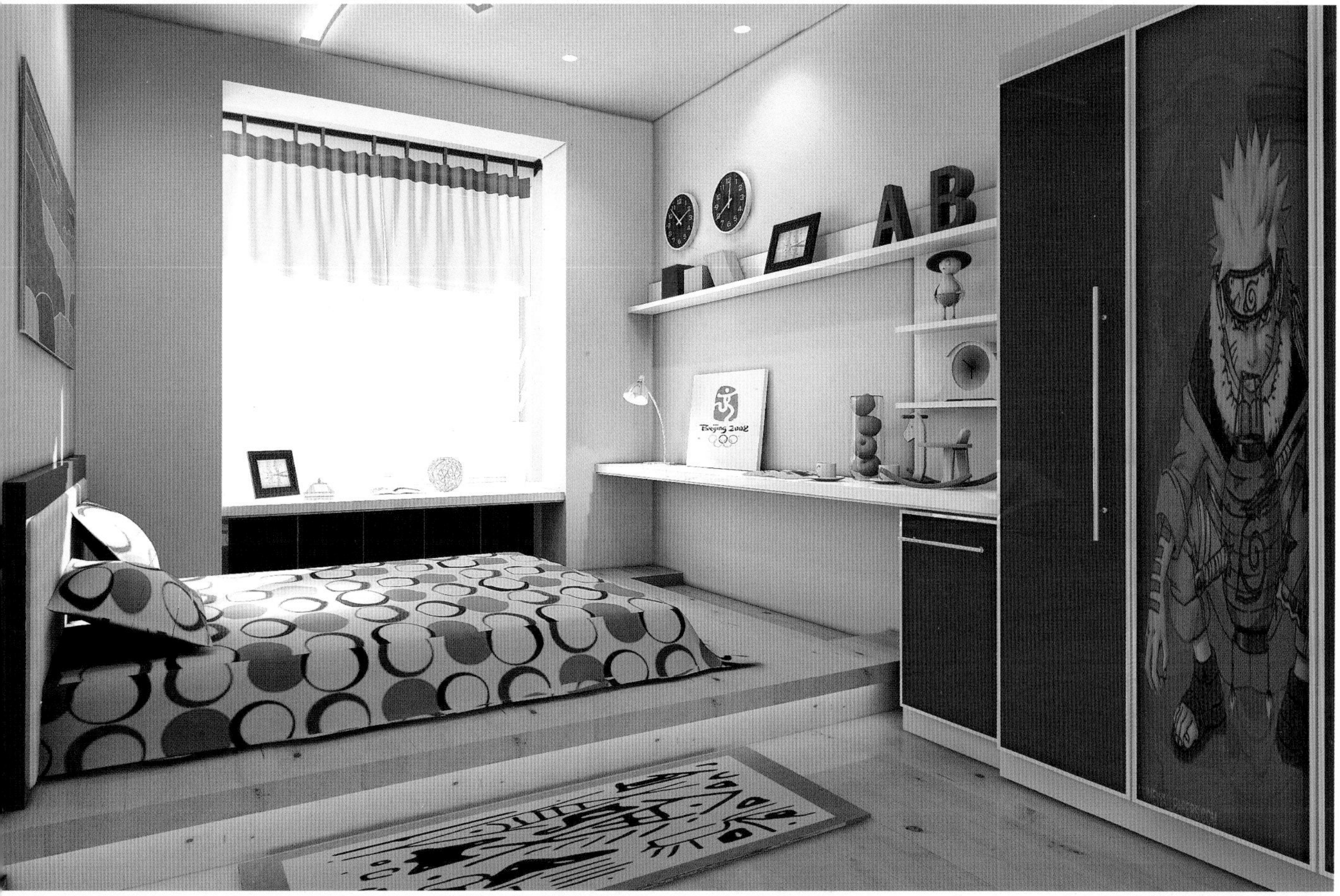
AB
Beijing 2008

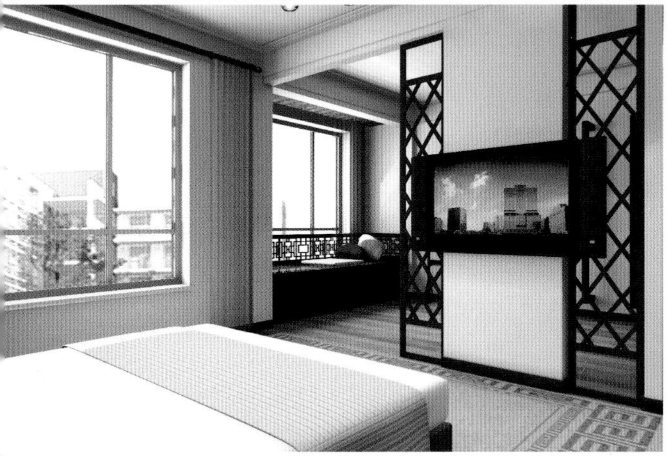

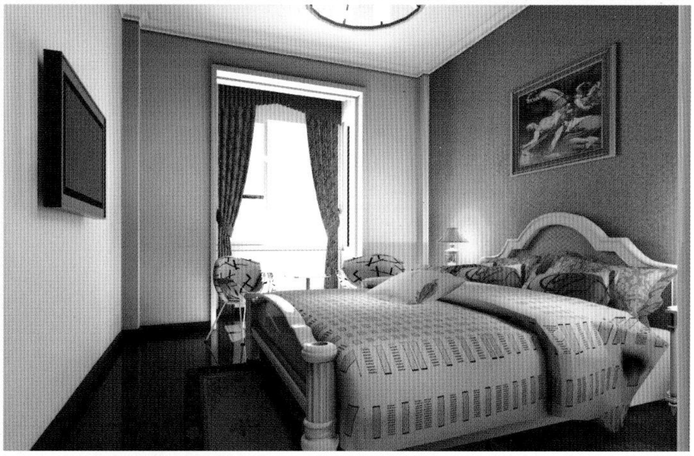

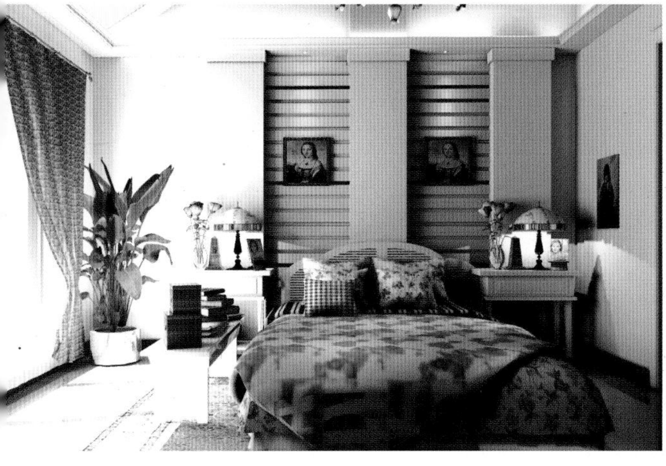

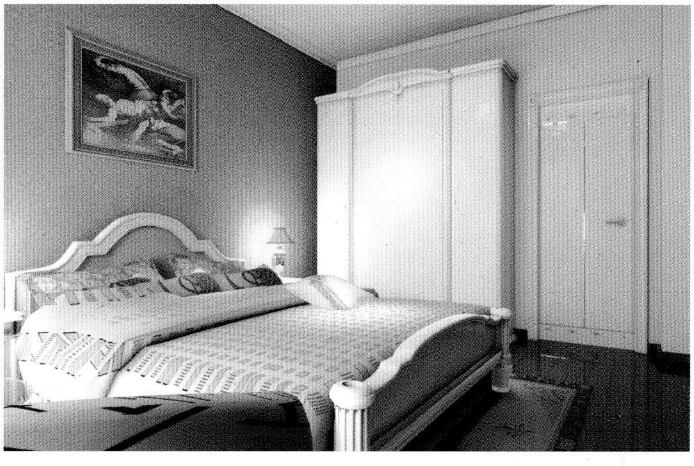

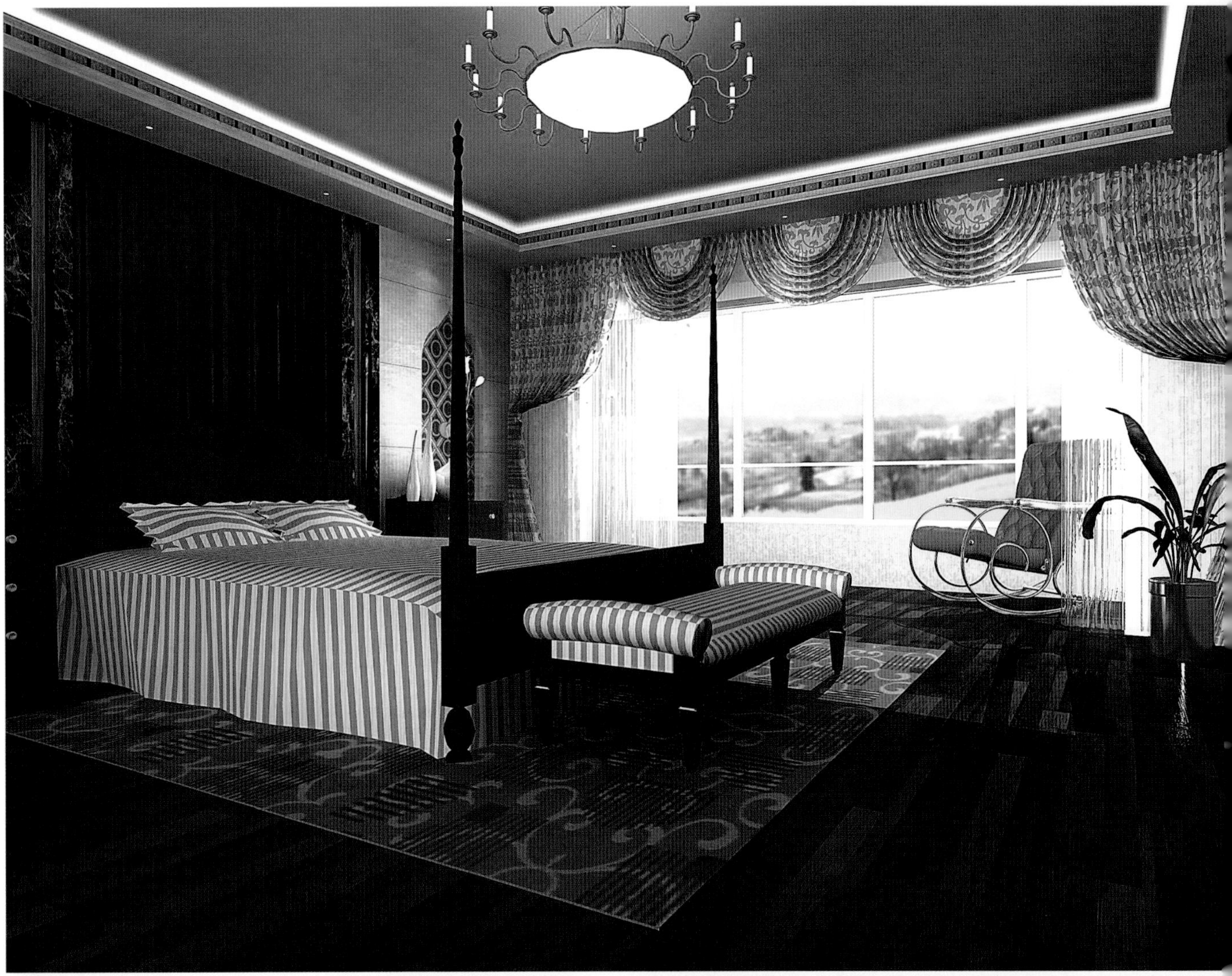

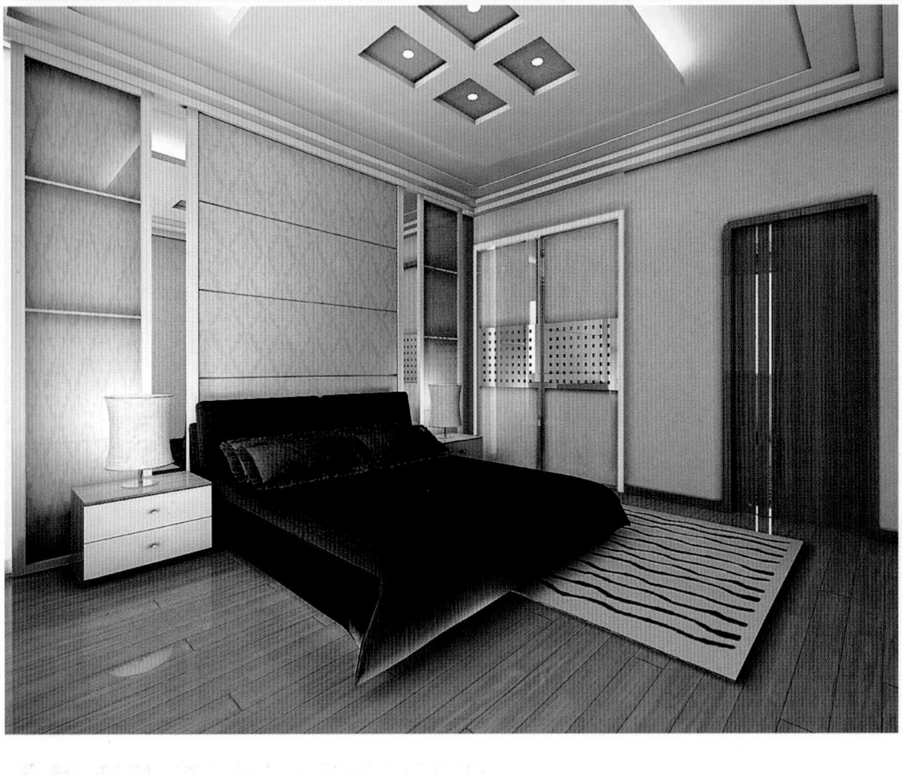

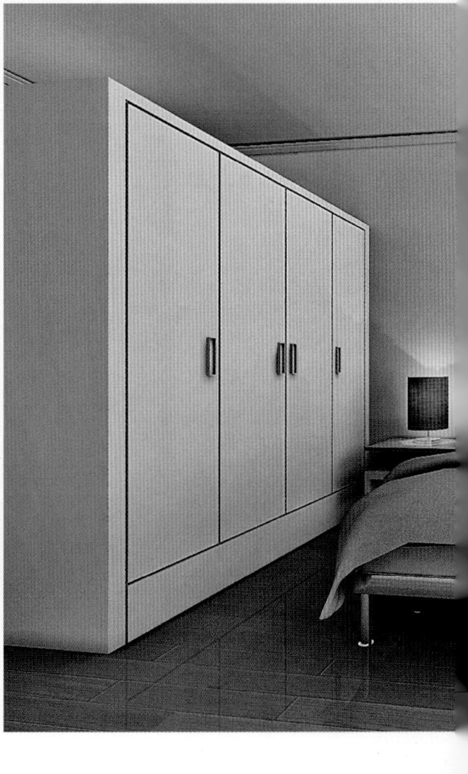

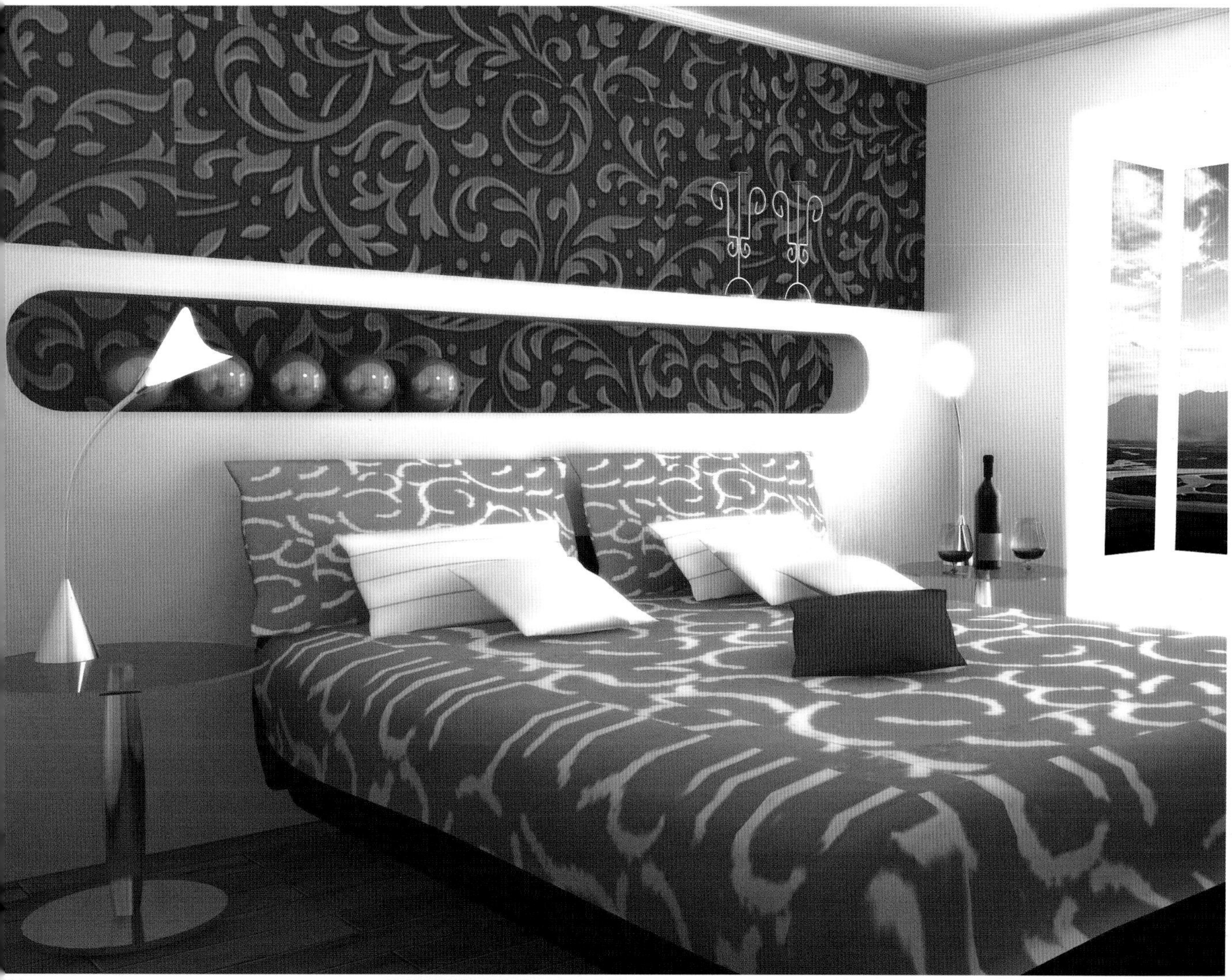

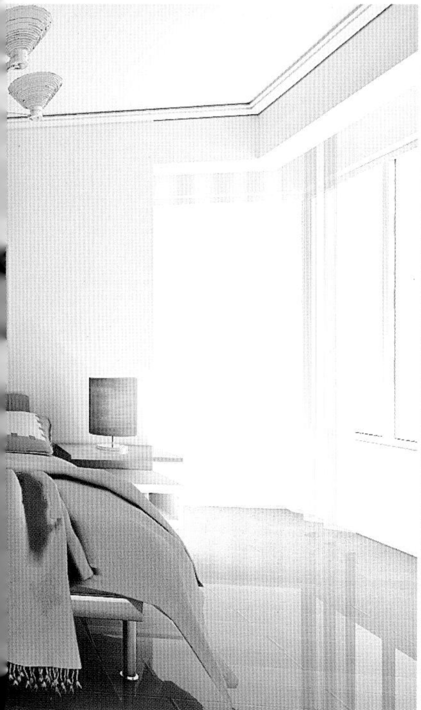

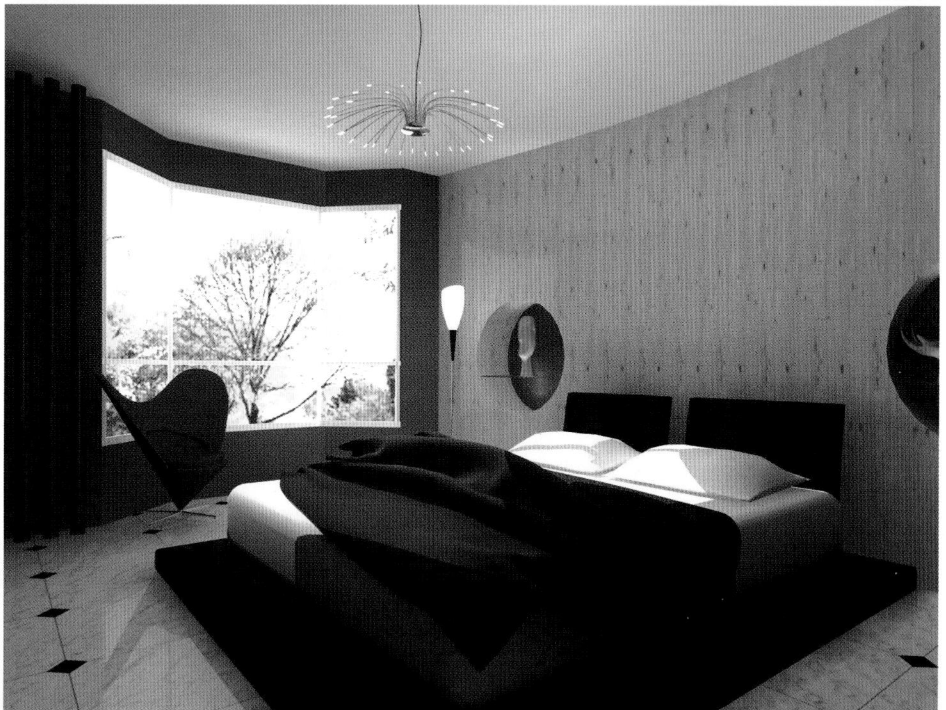

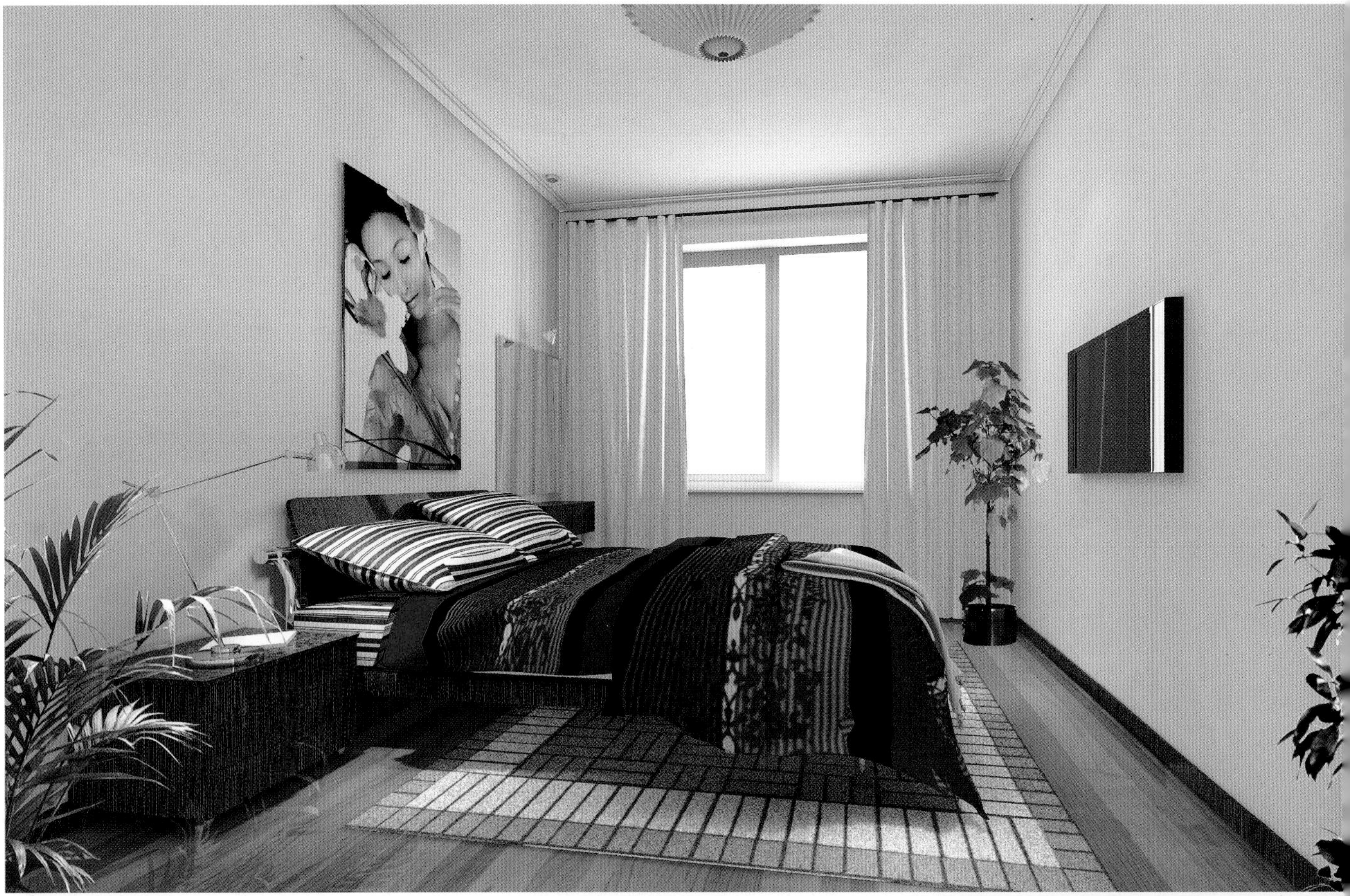

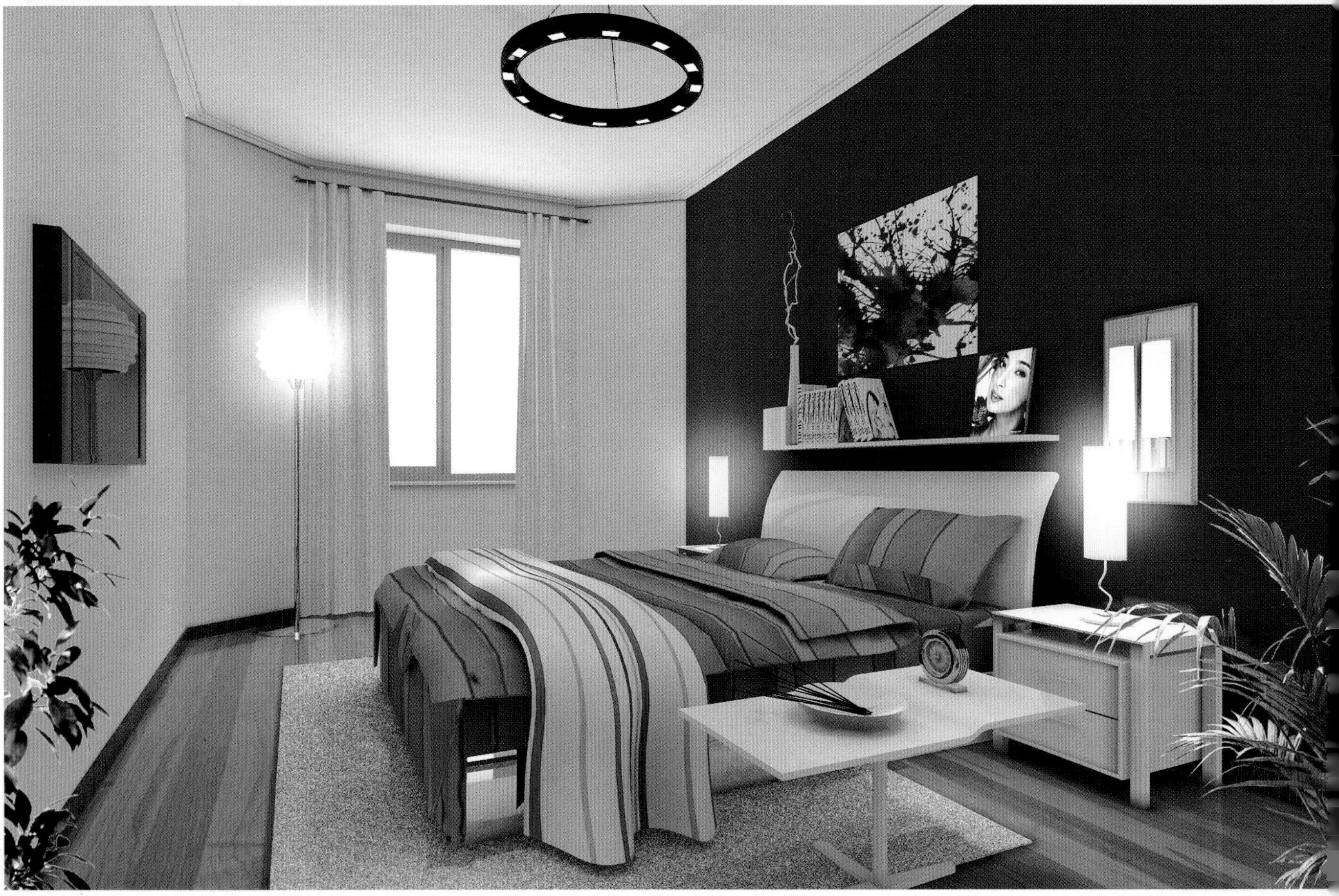

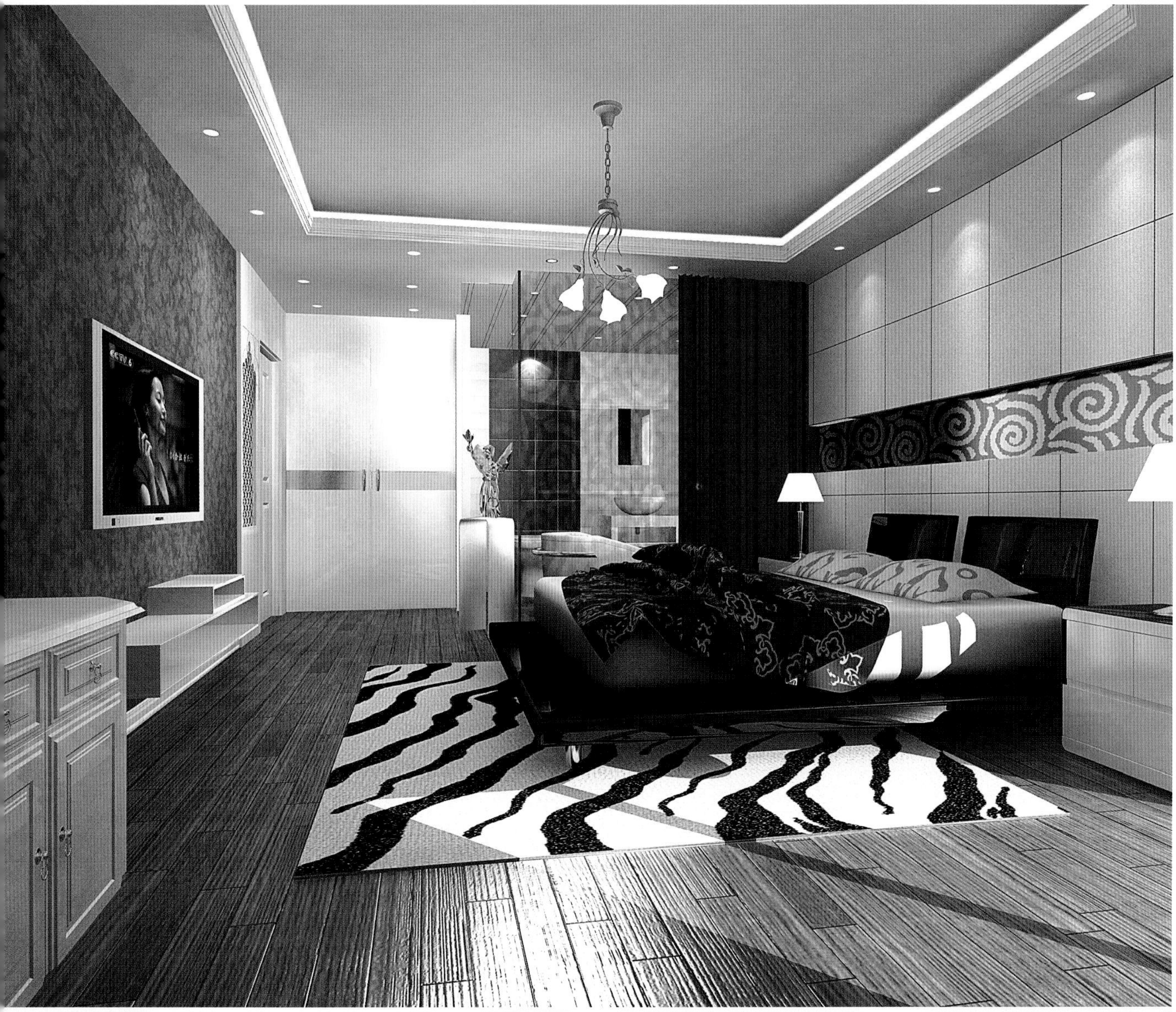

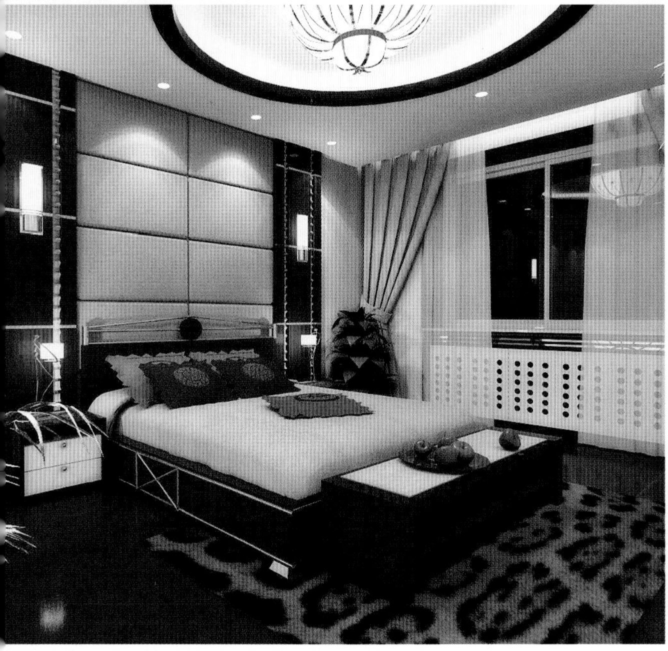

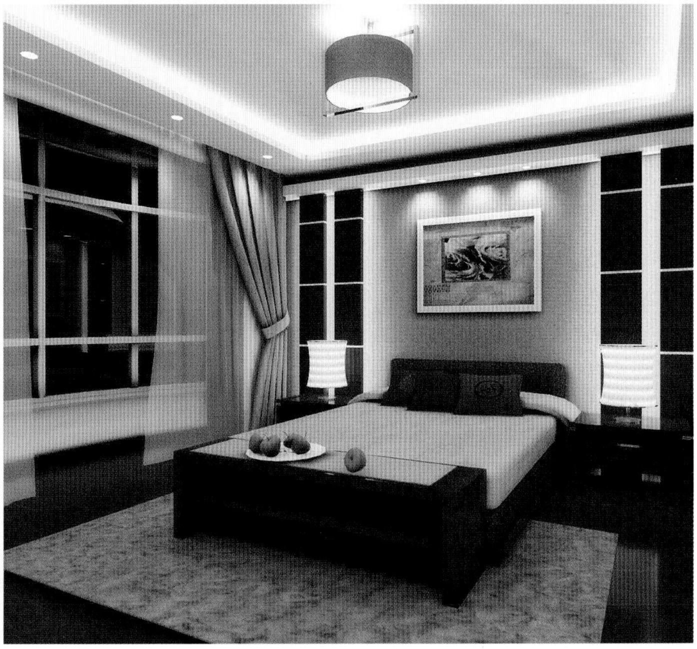

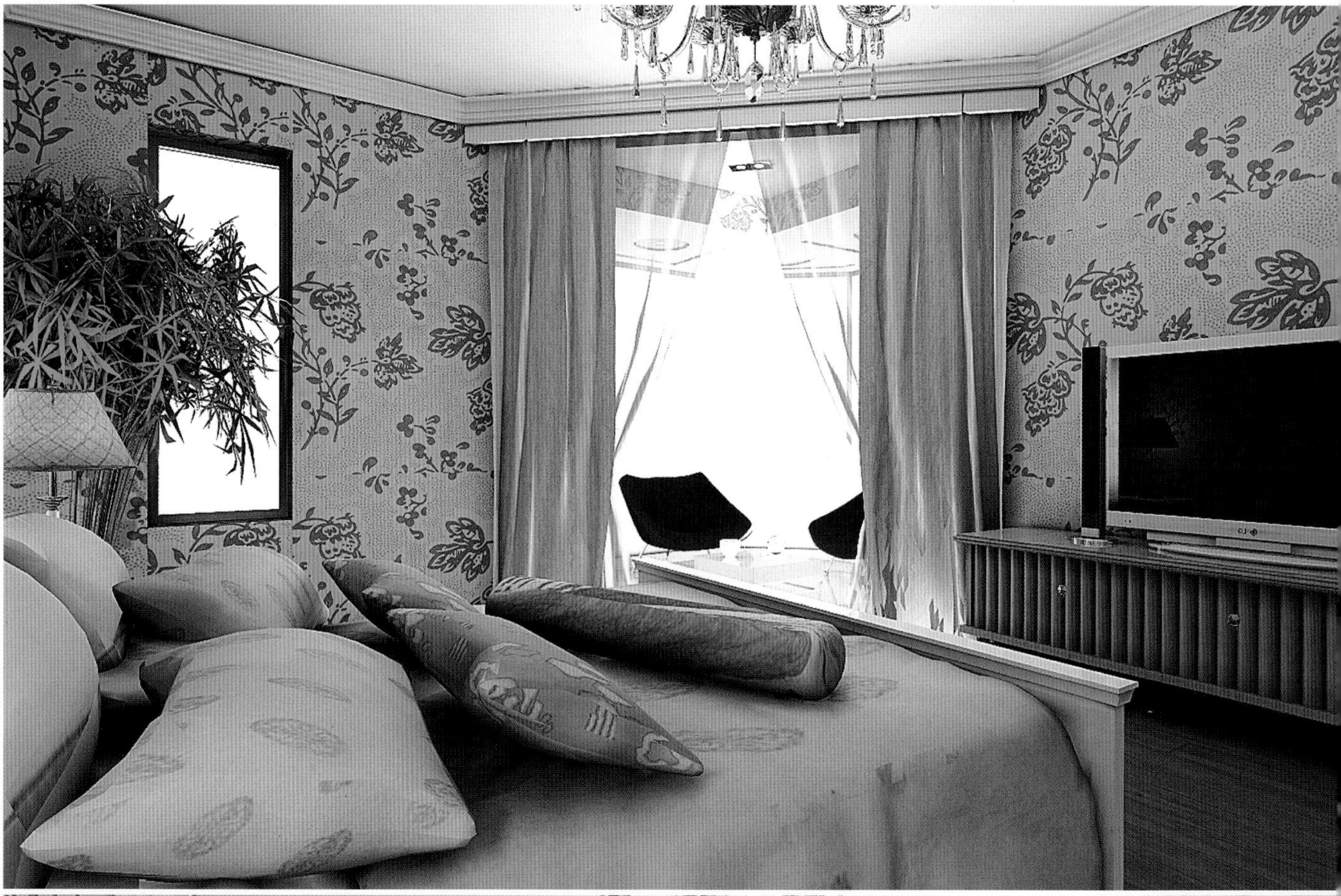

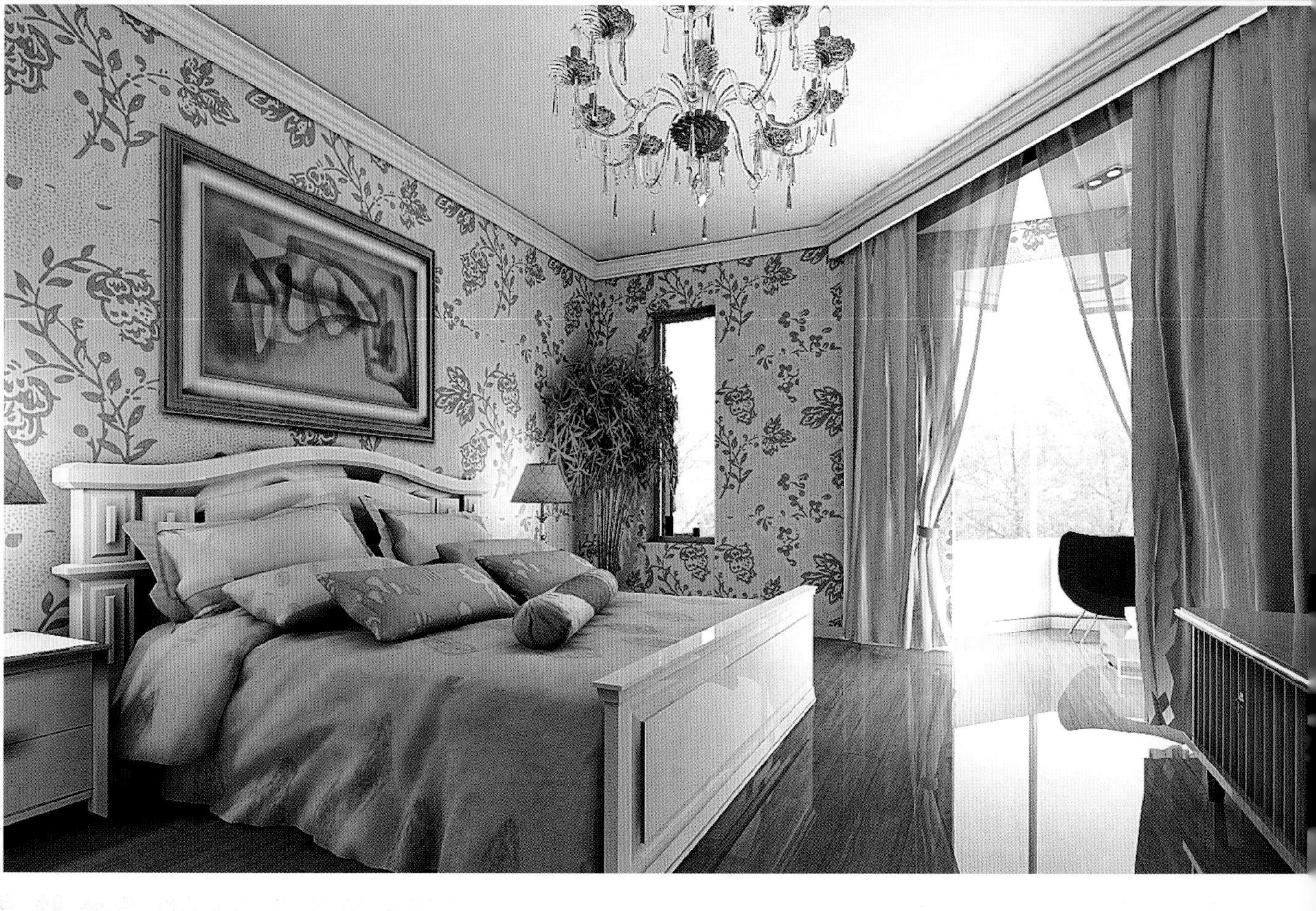

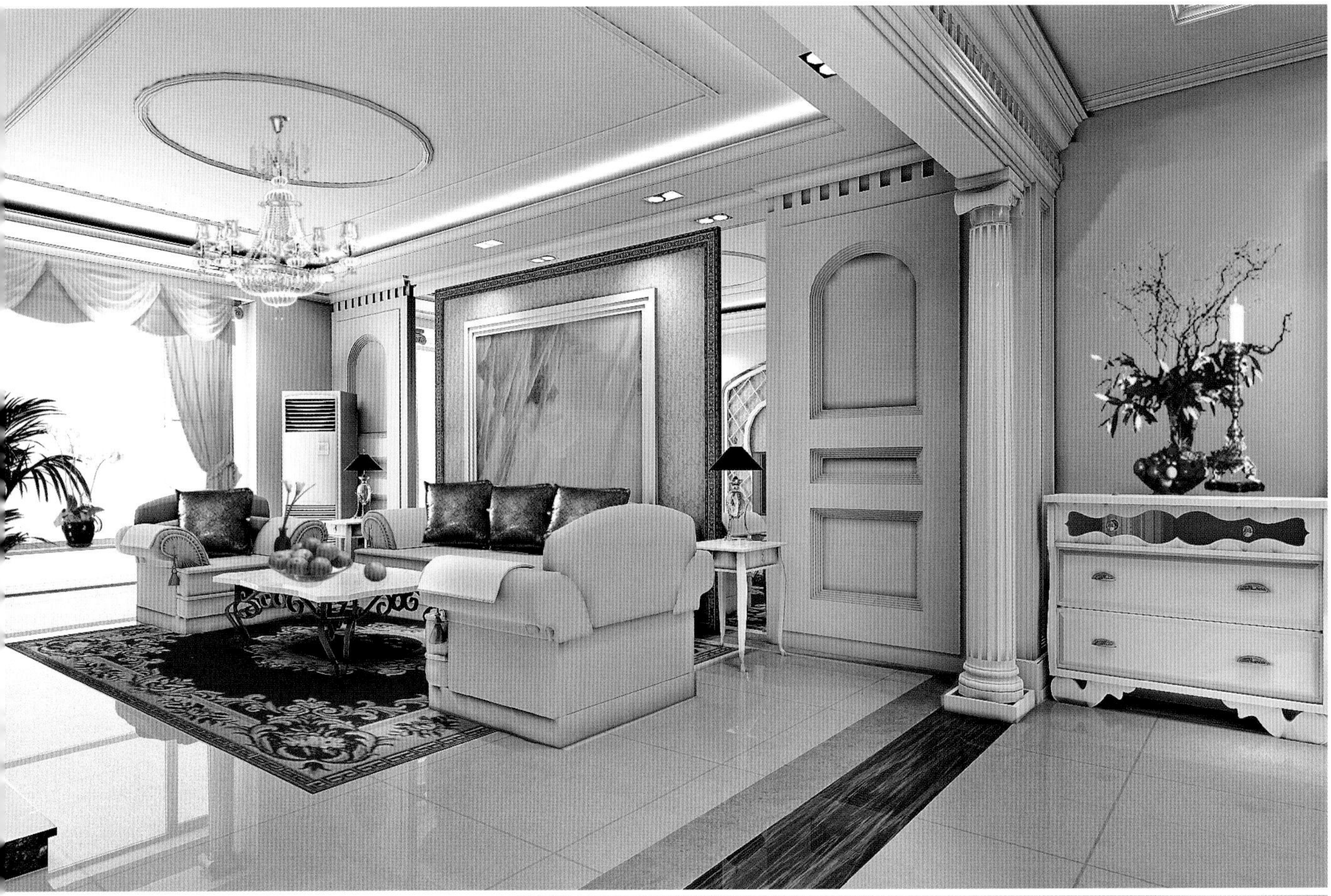

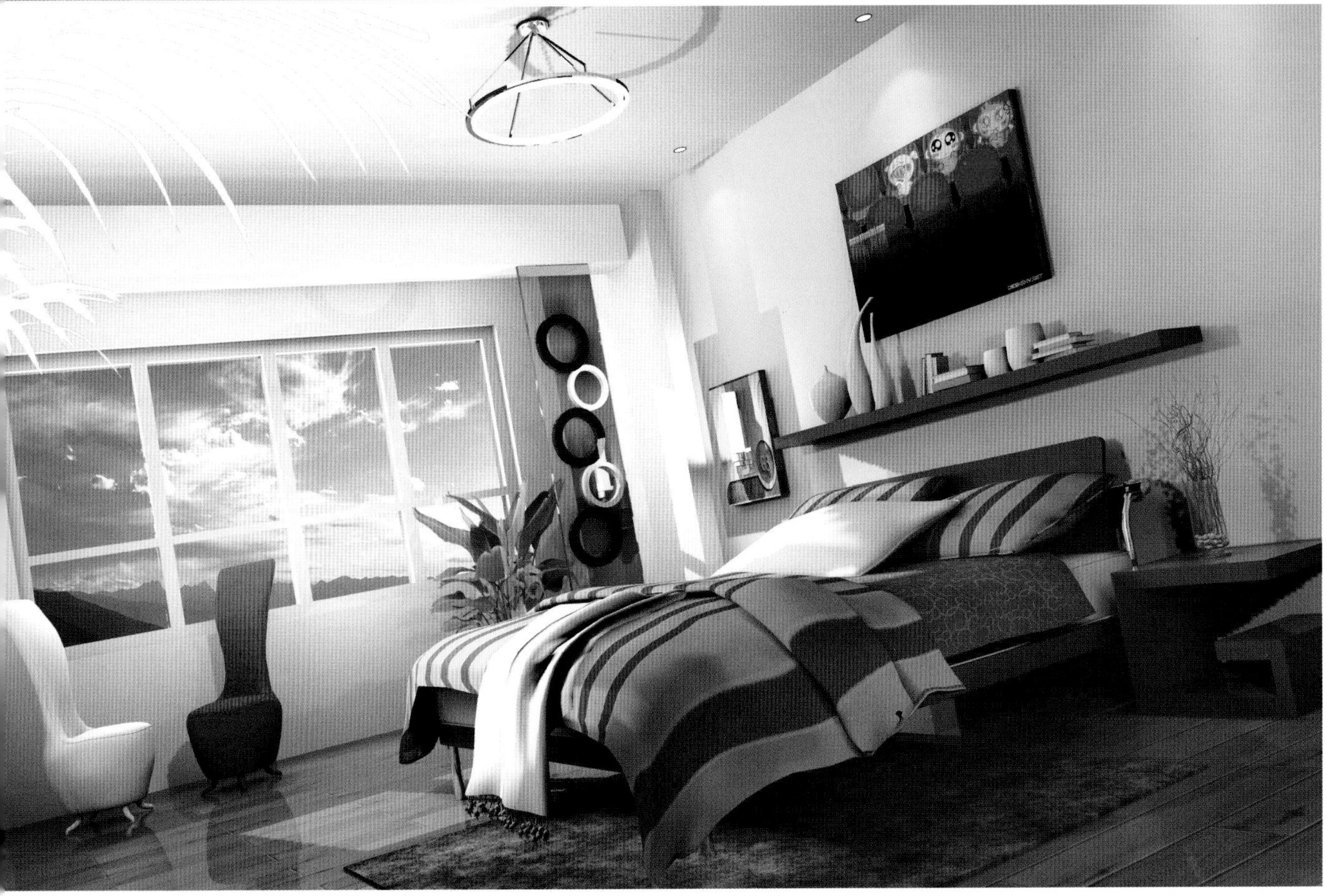

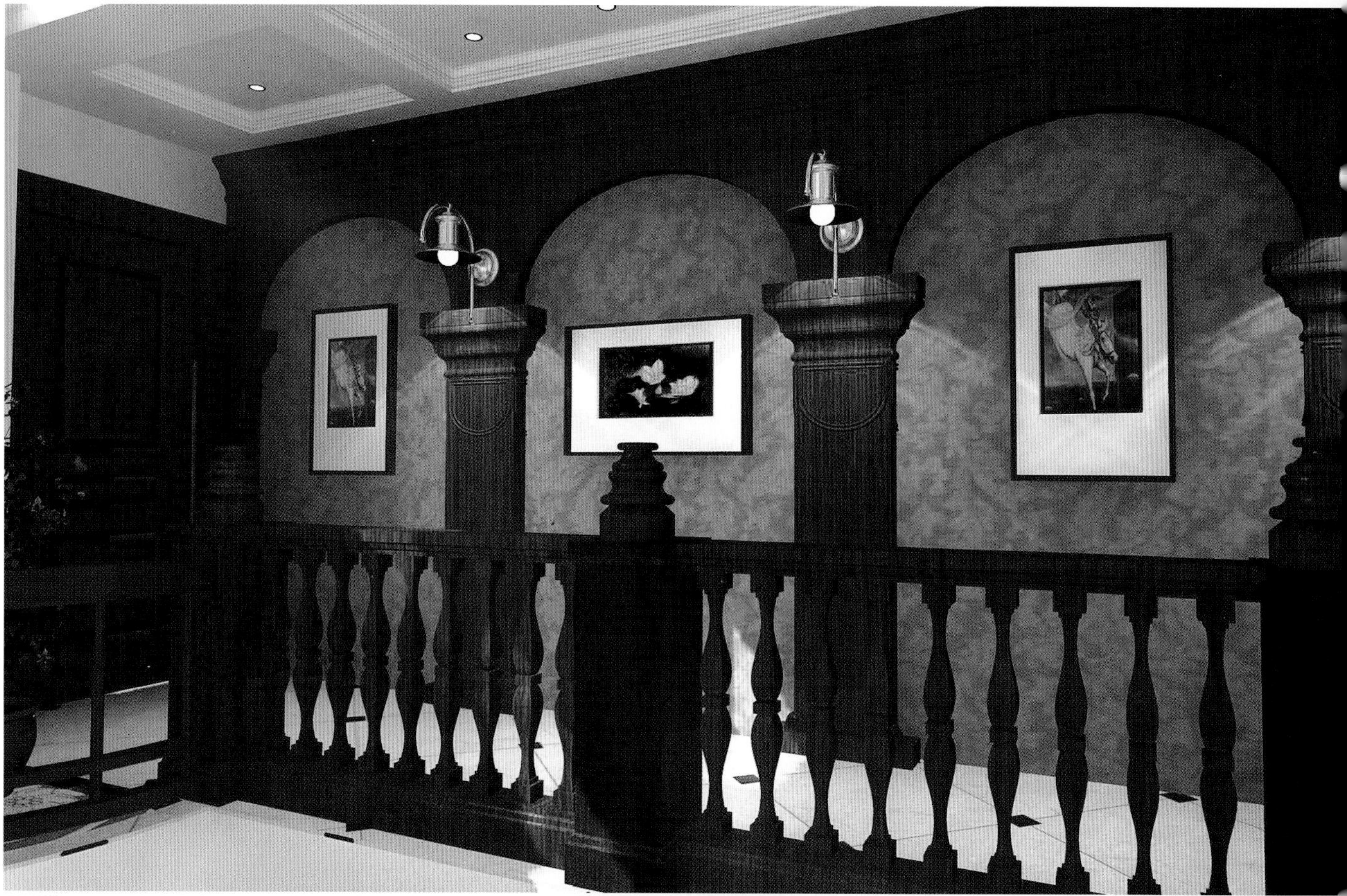

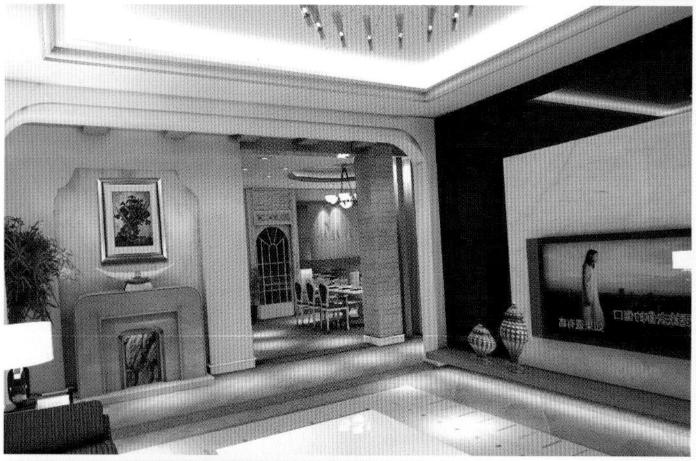

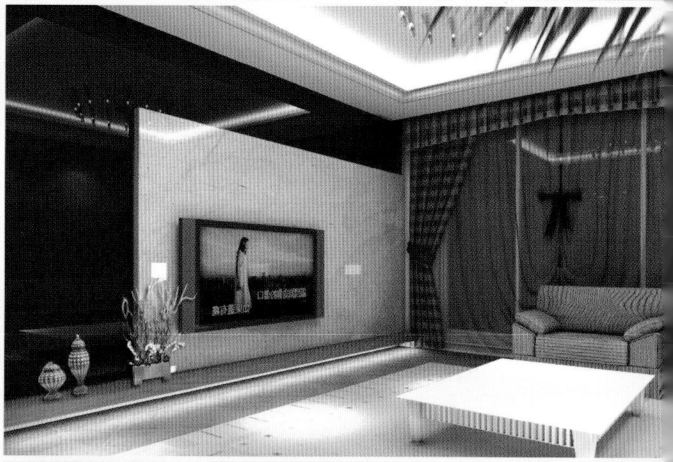

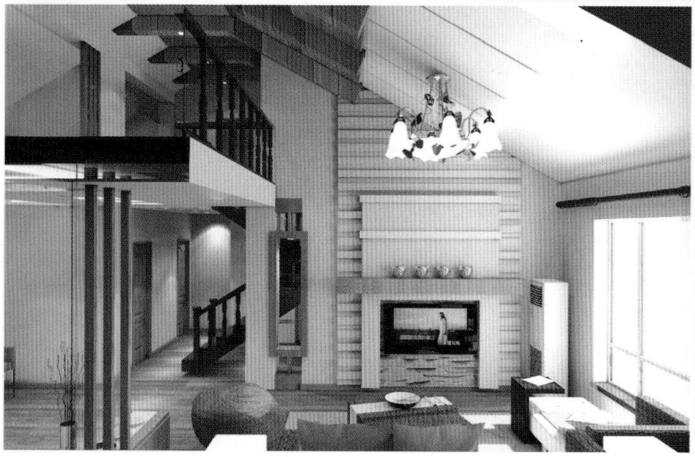

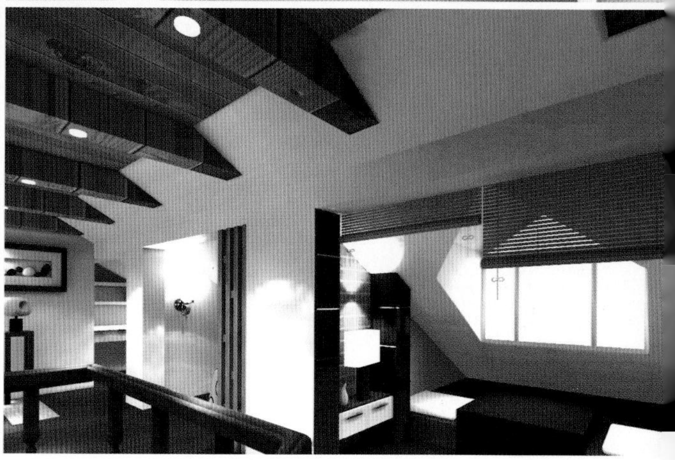

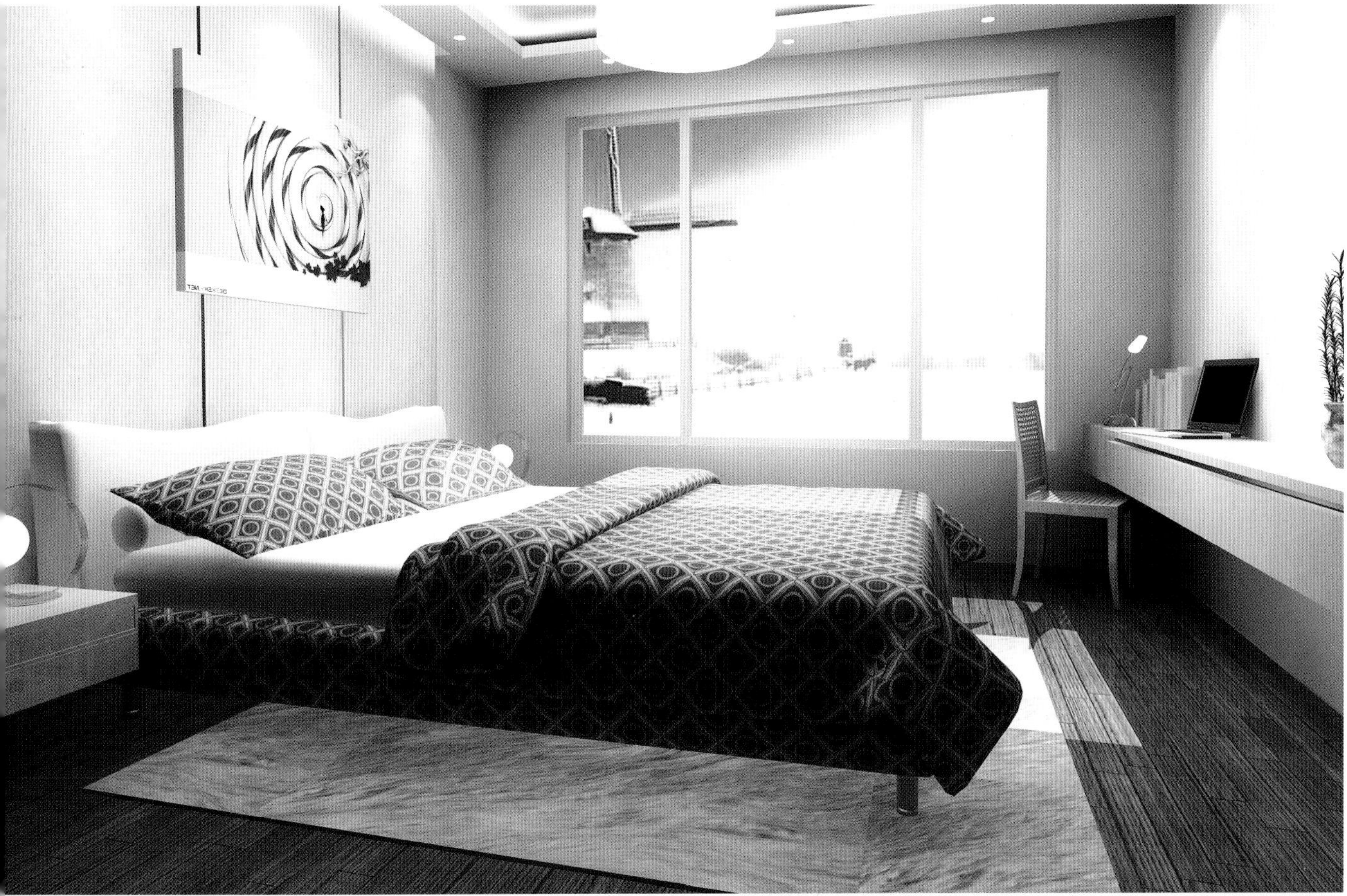

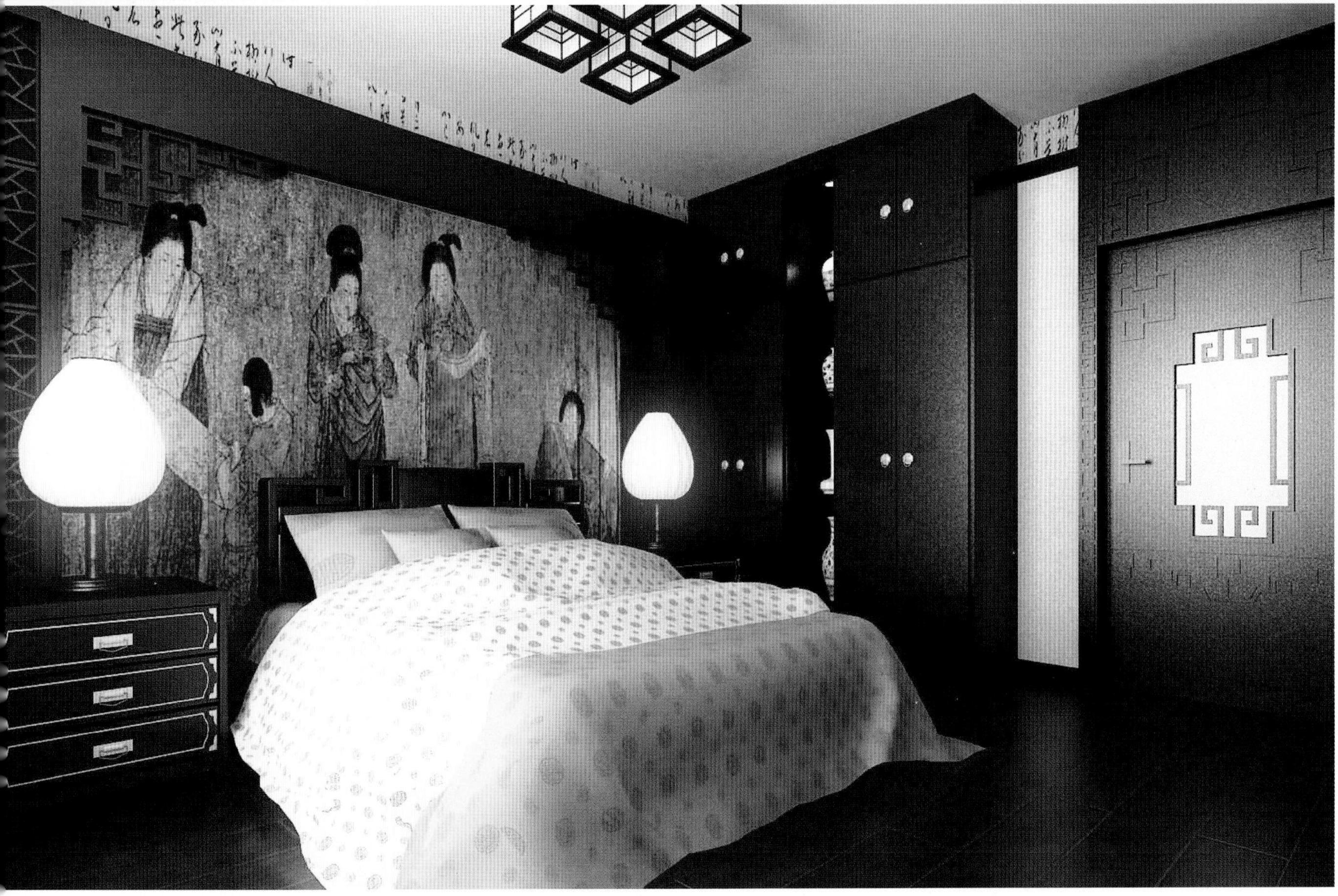

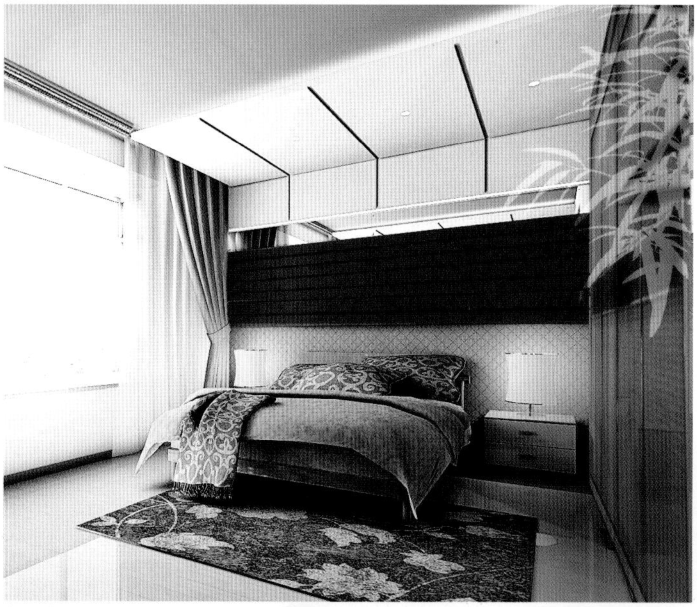
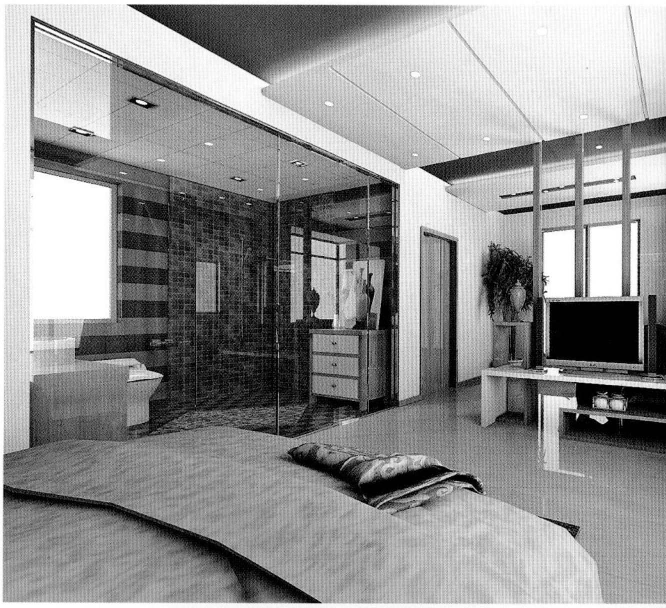
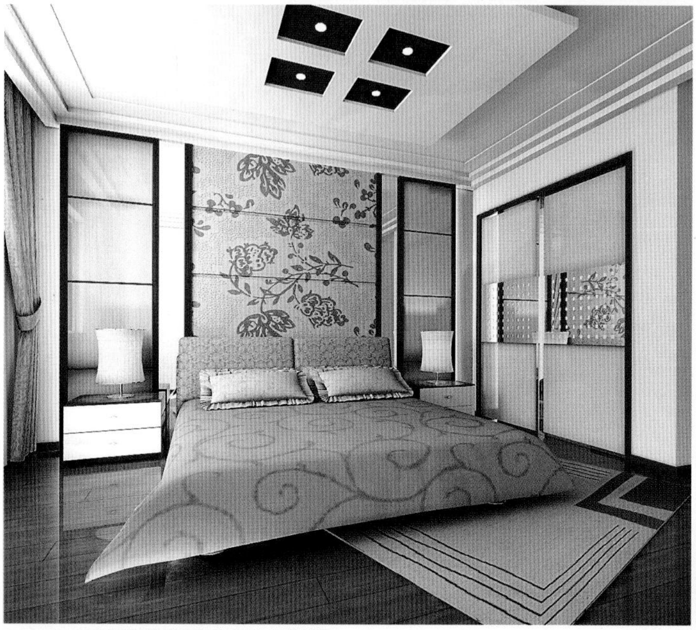
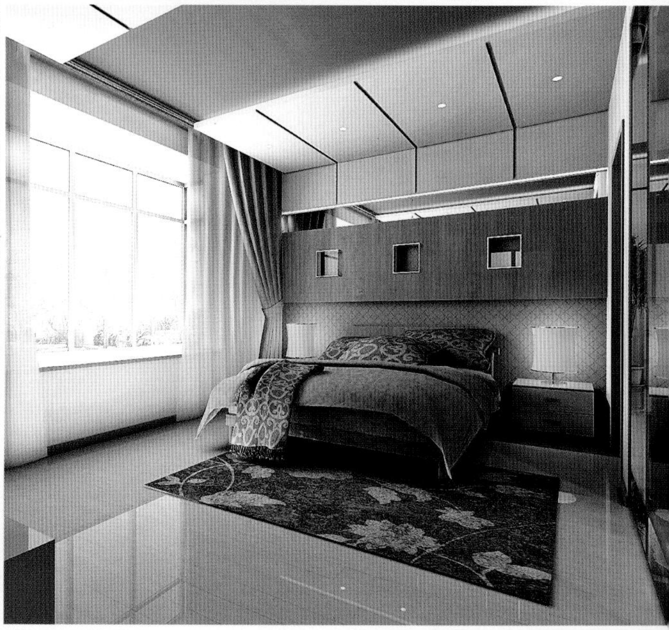
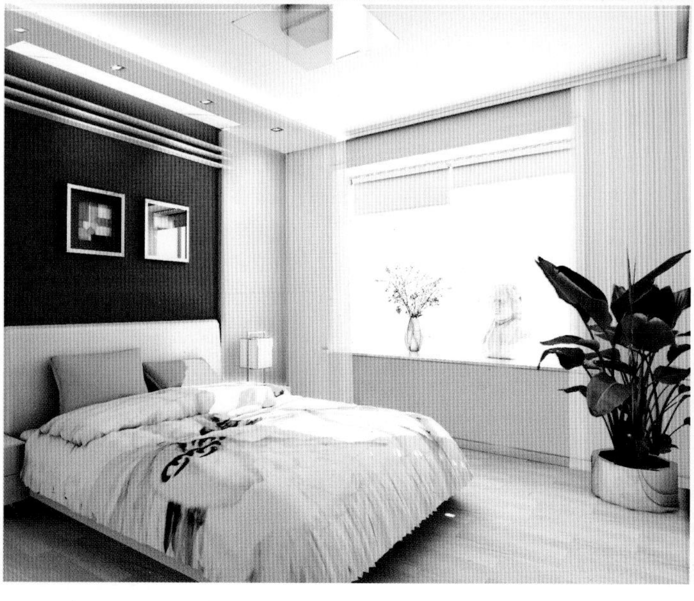
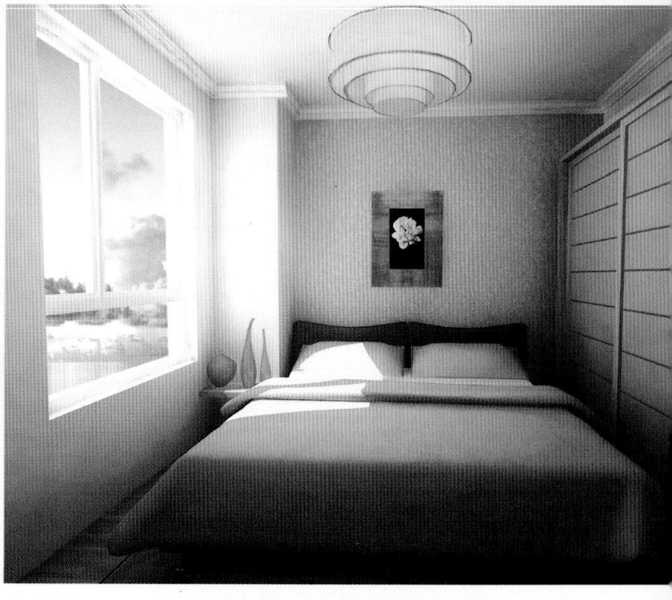

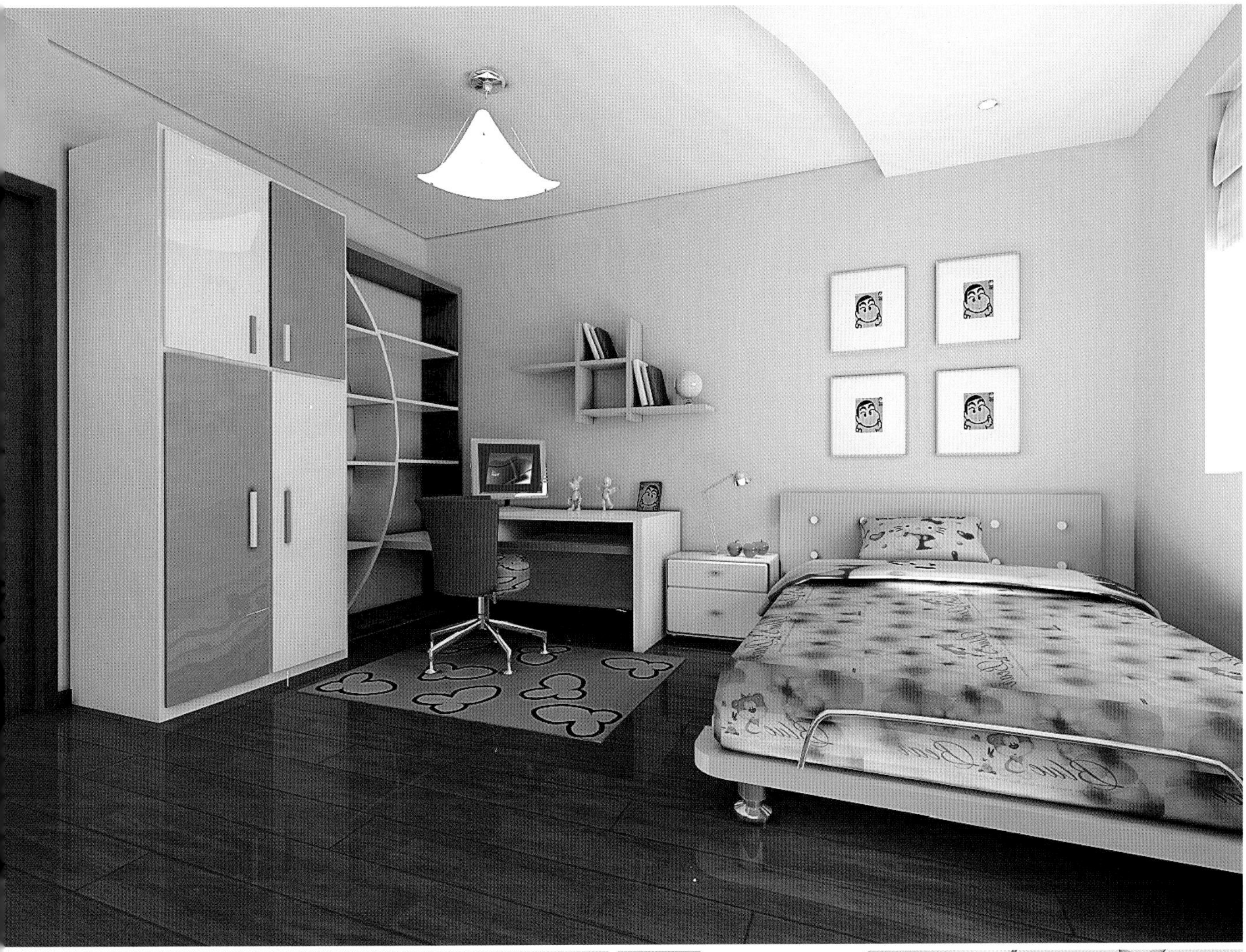

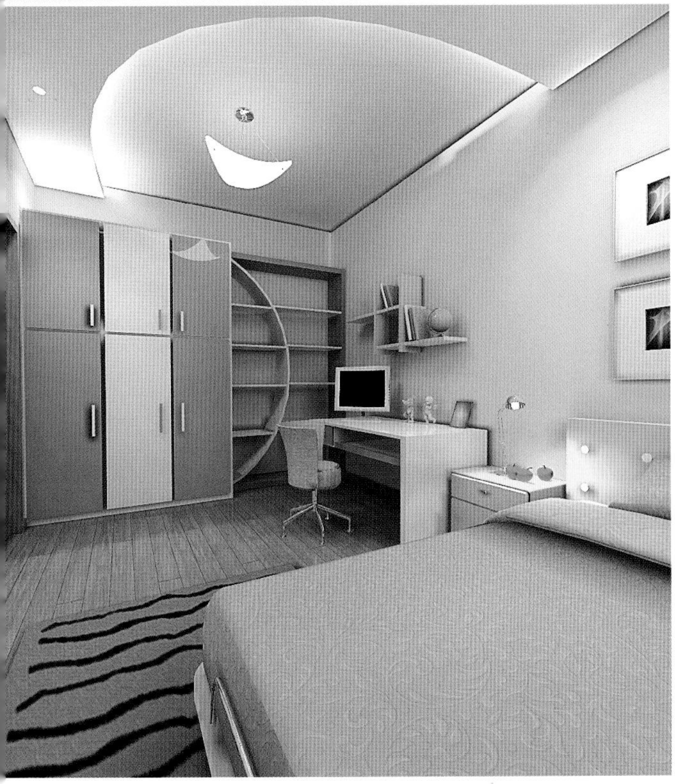

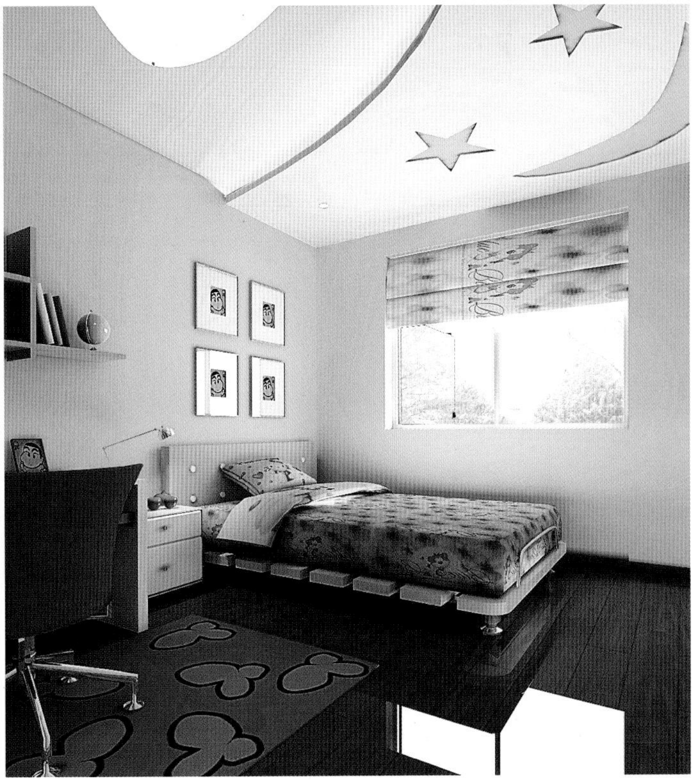

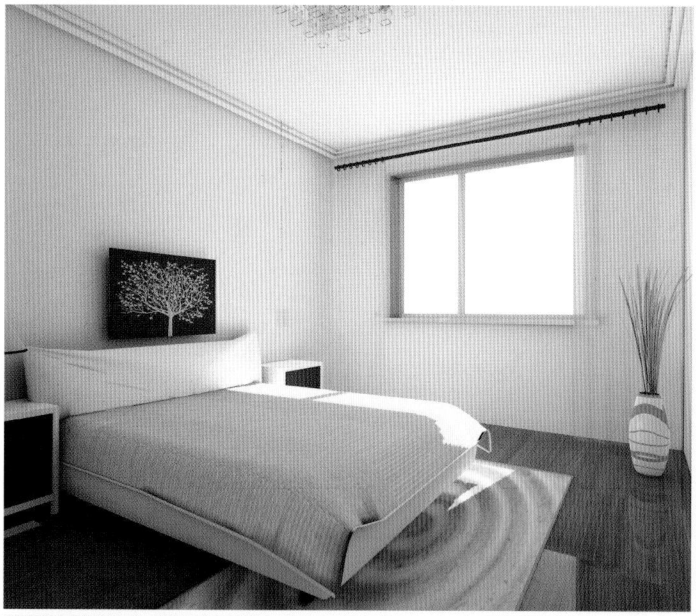

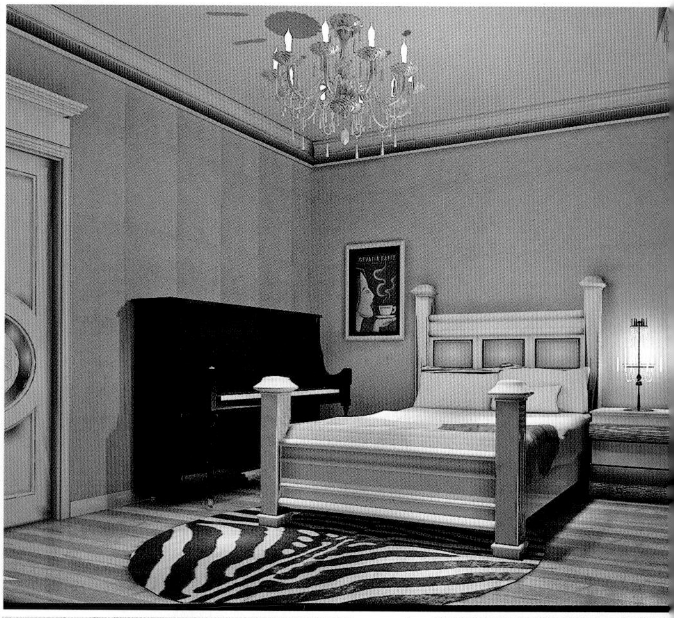

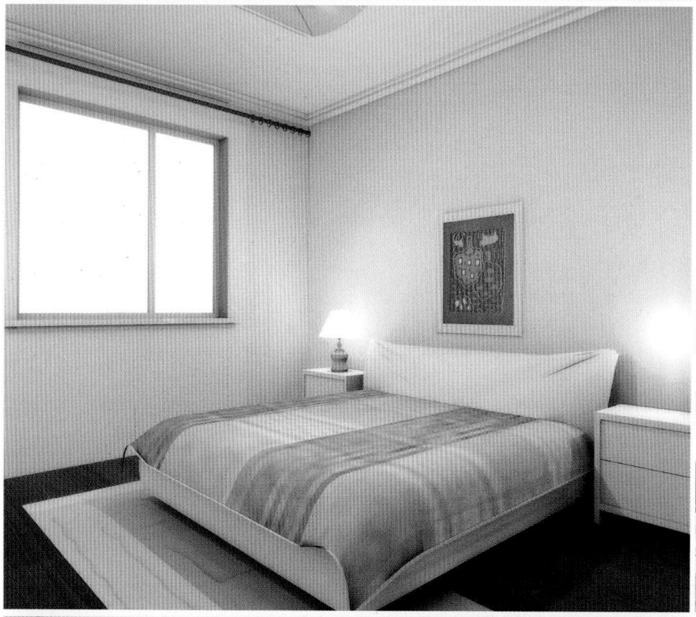

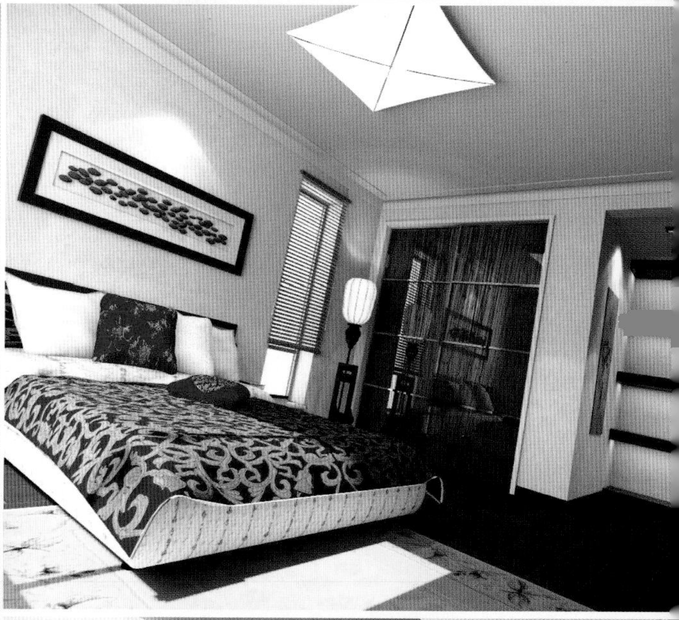

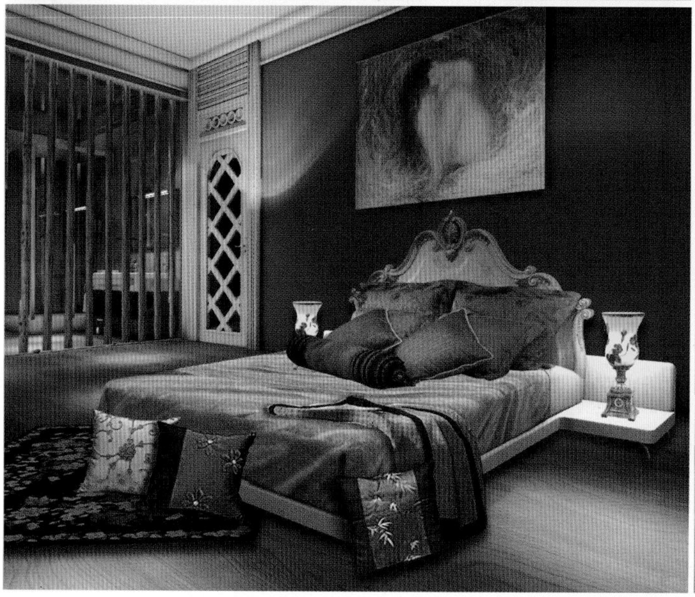

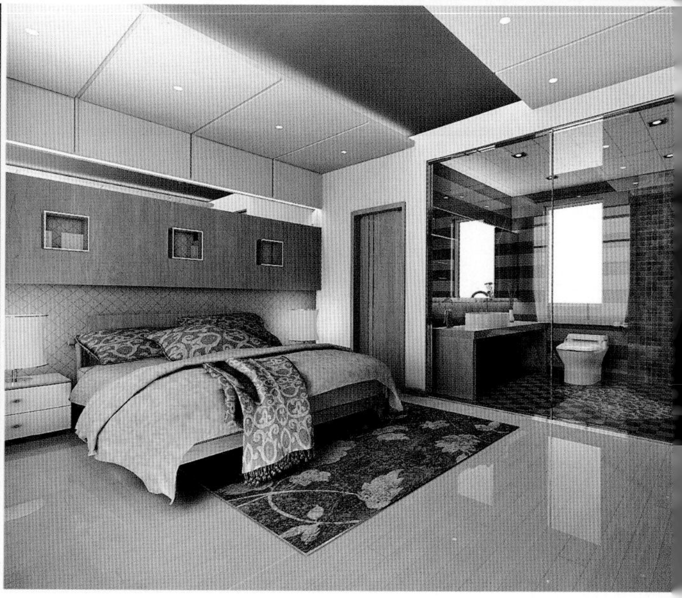

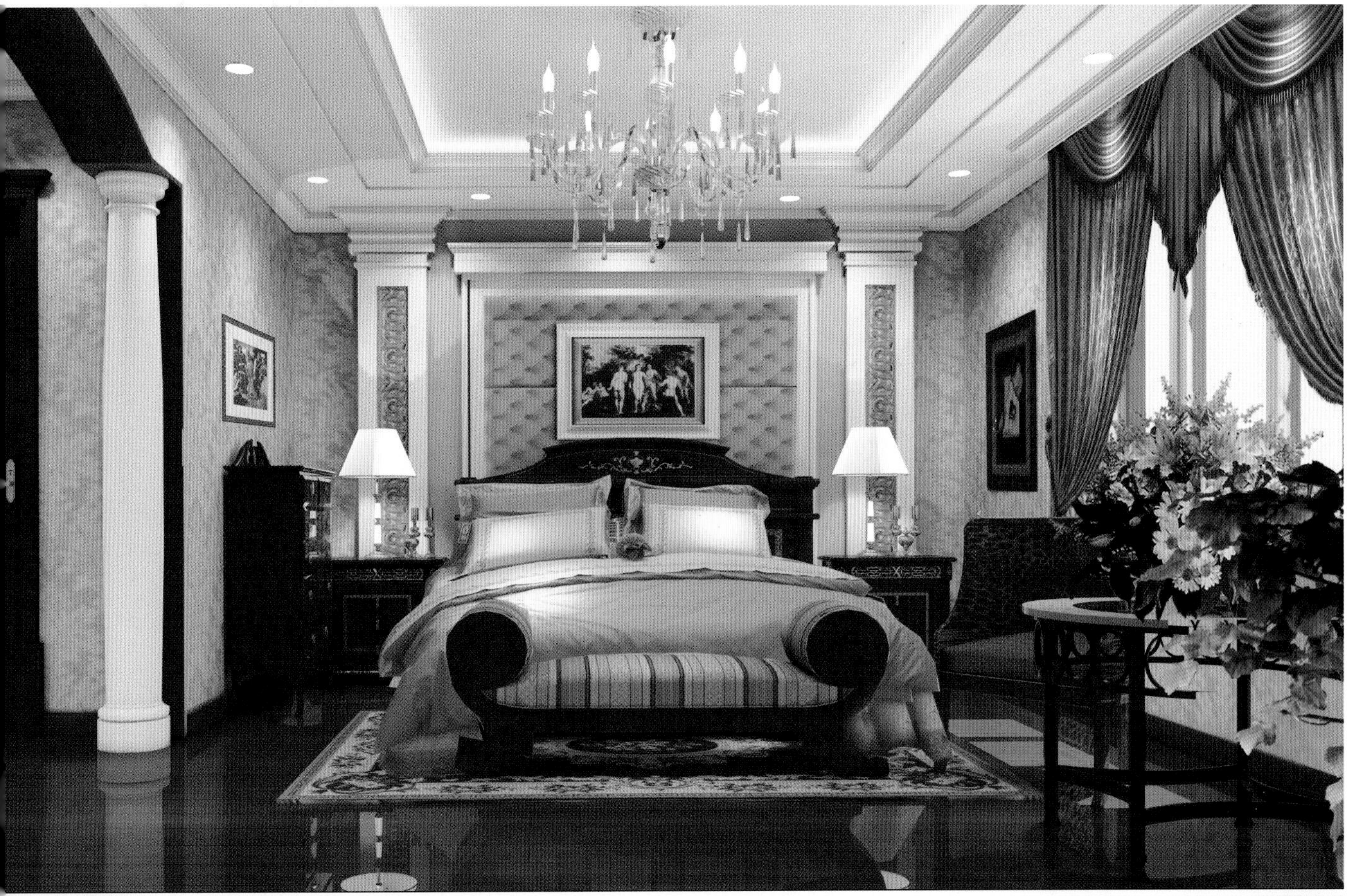

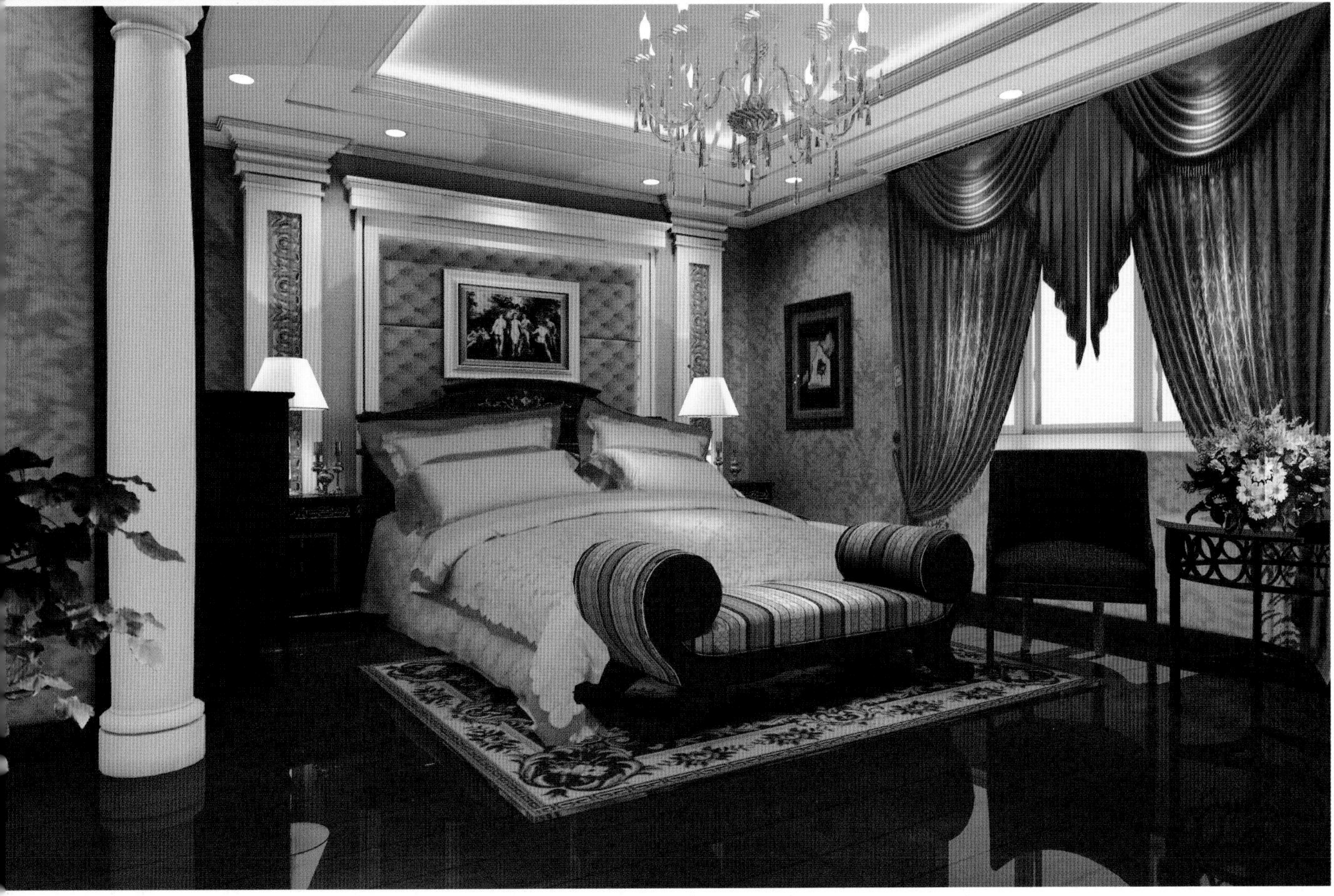

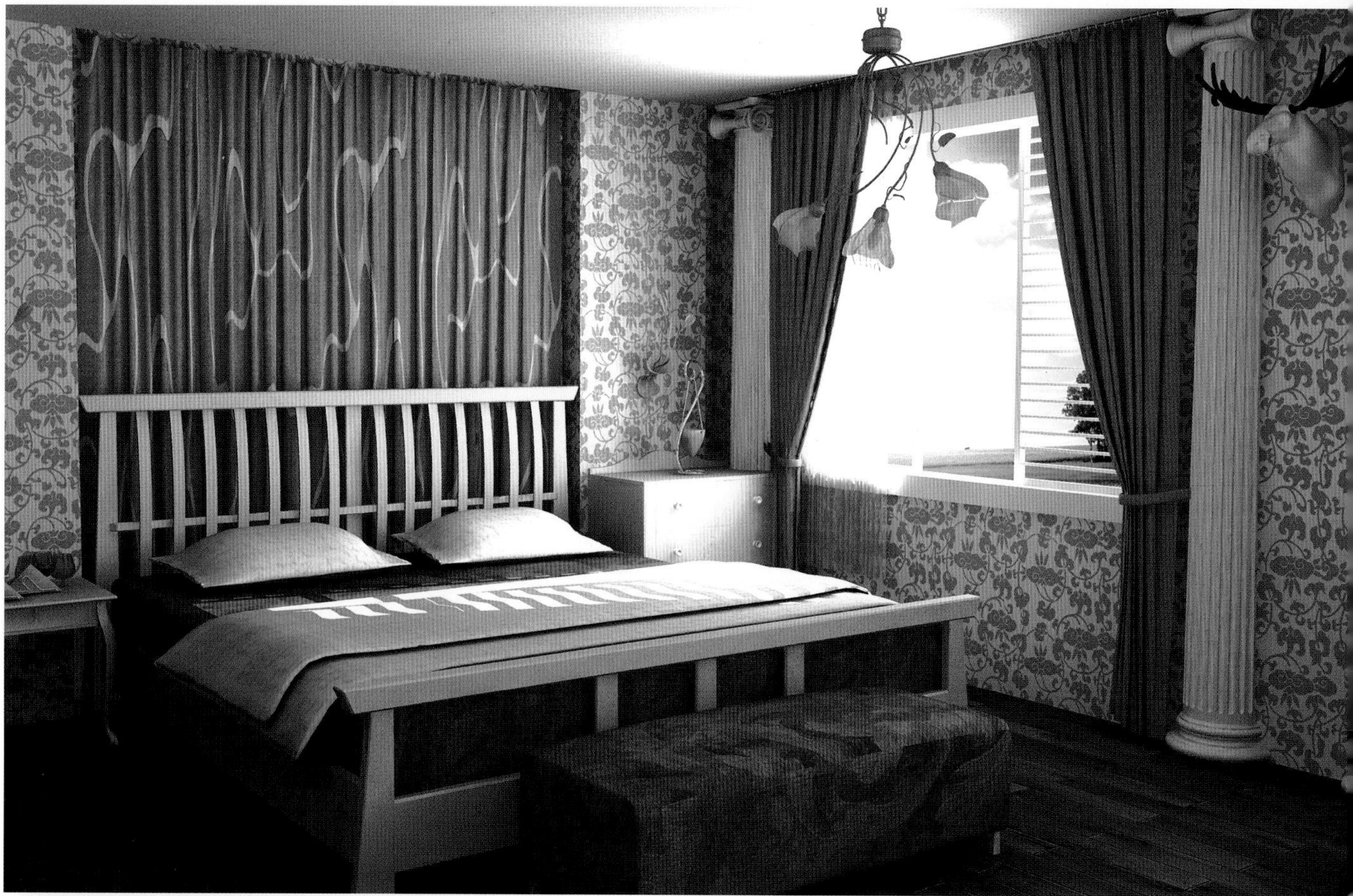

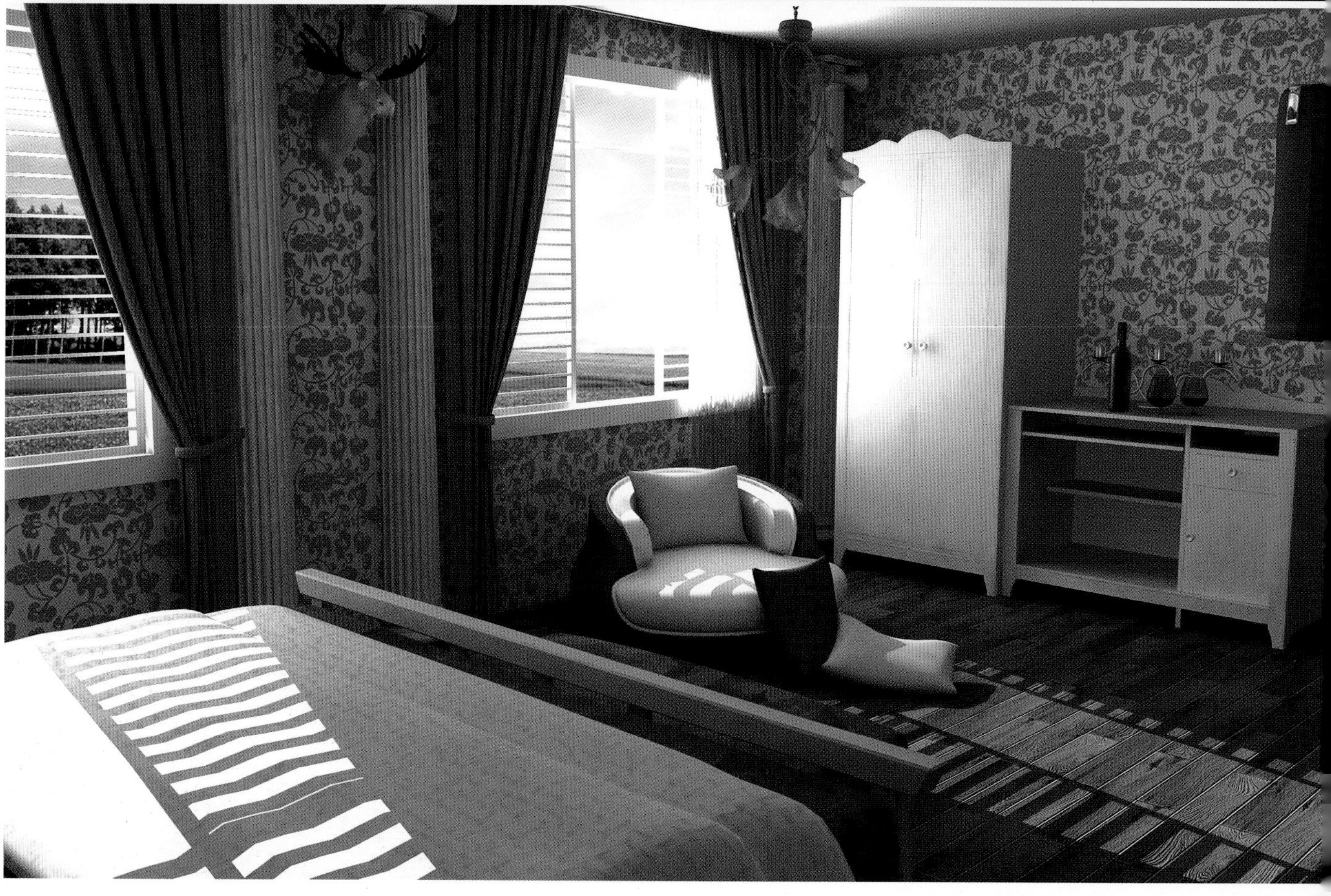

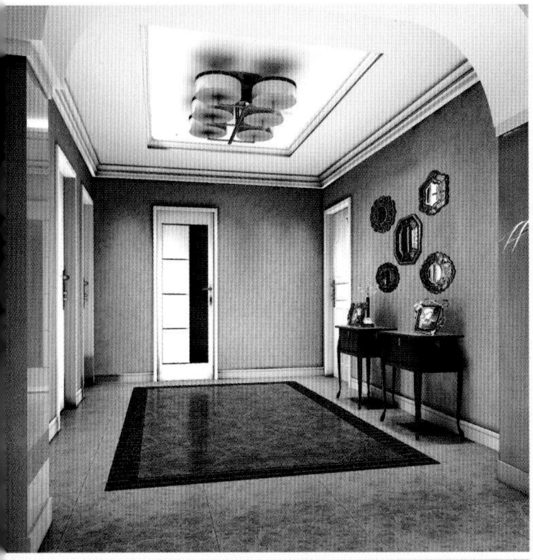

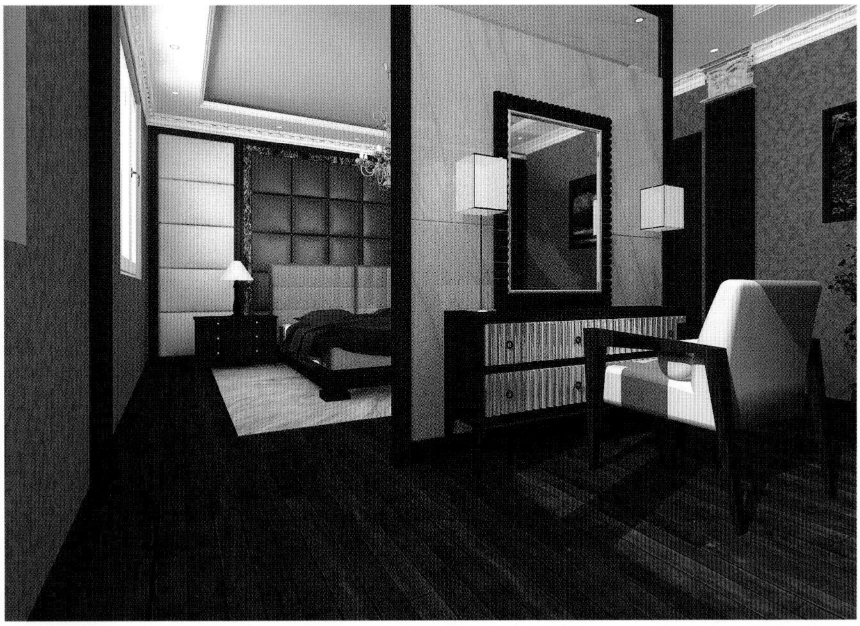

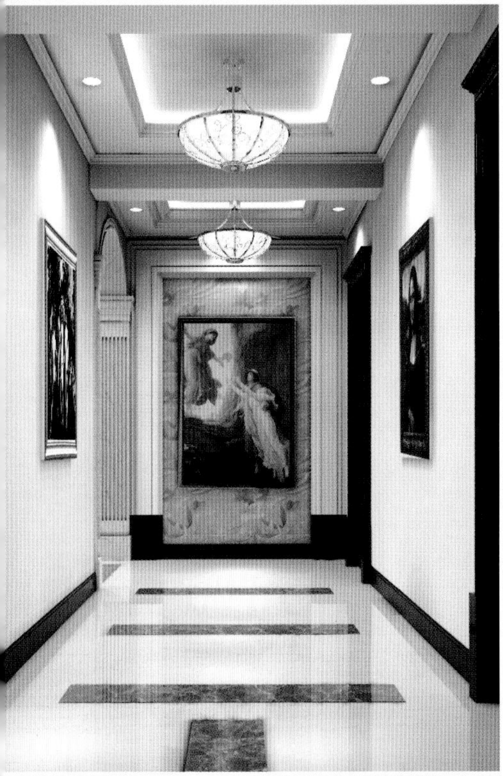

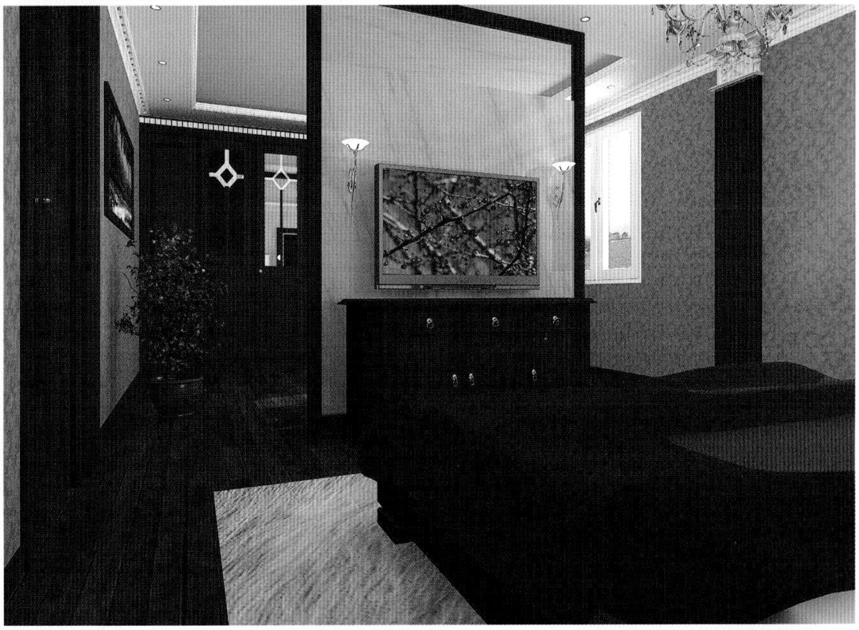

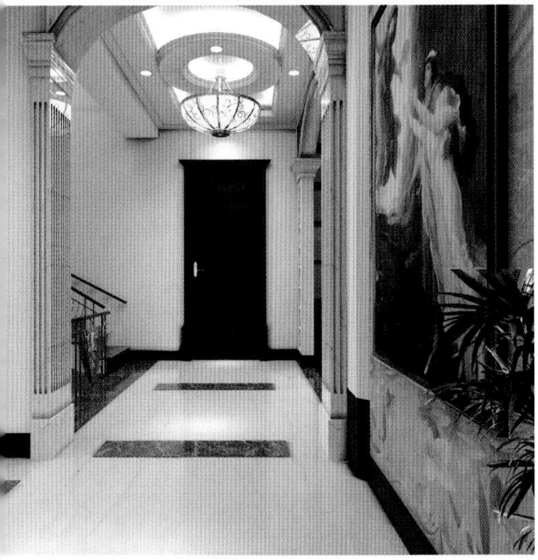

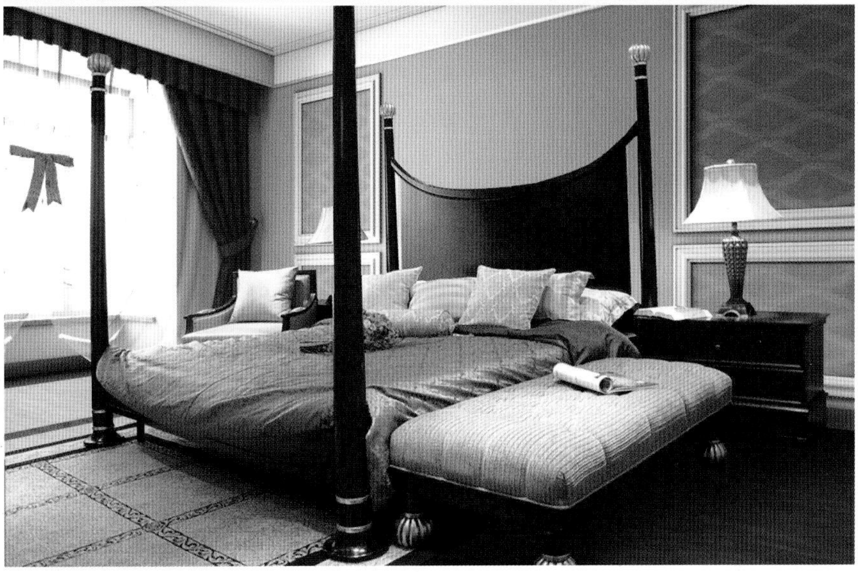

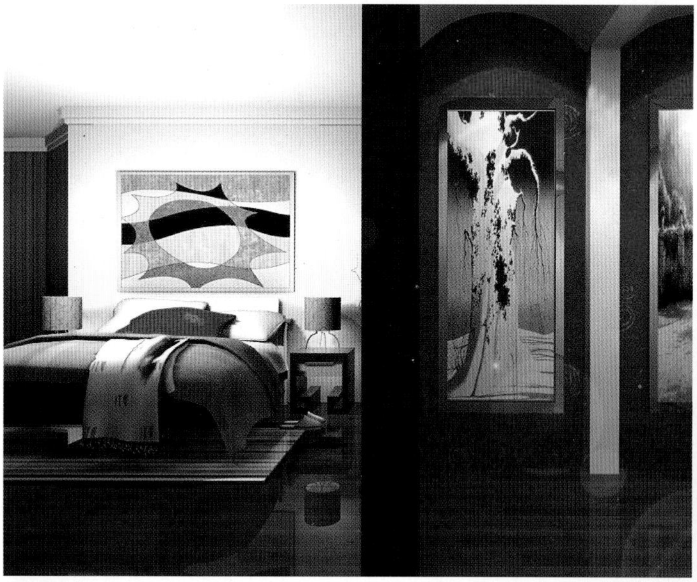

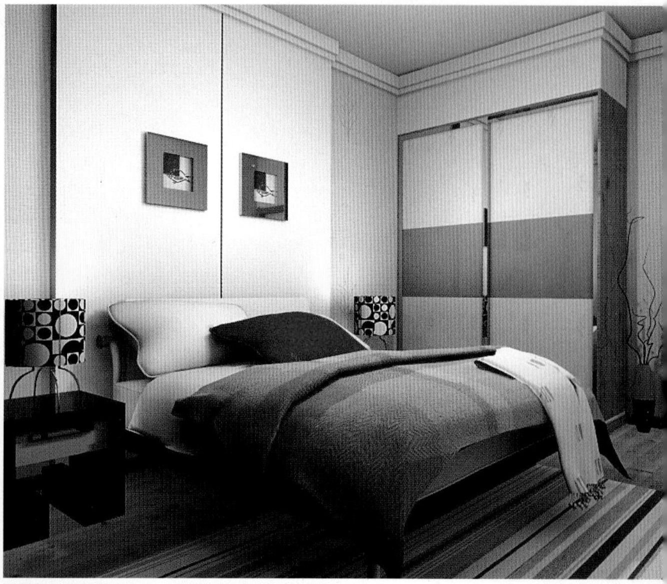

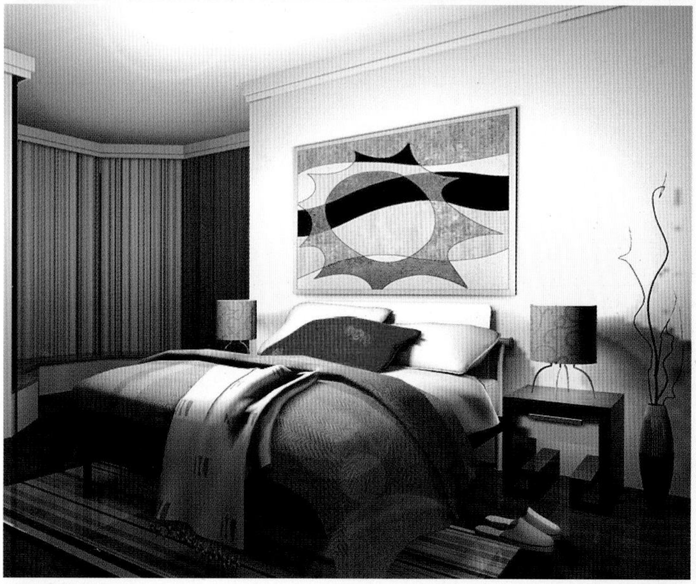

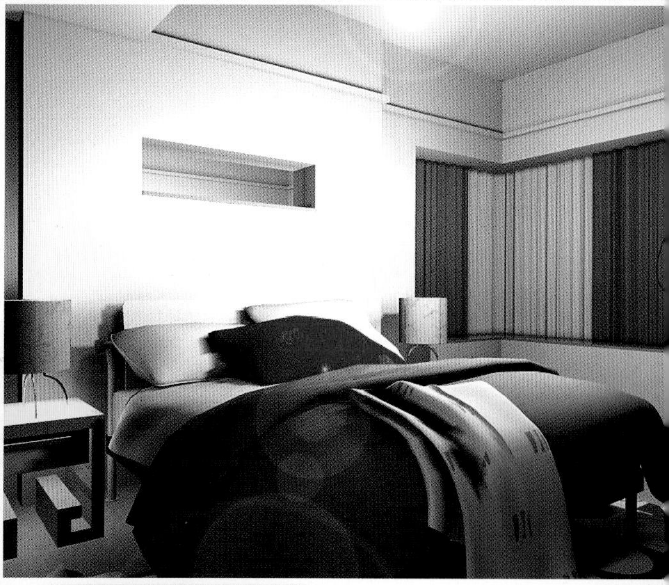

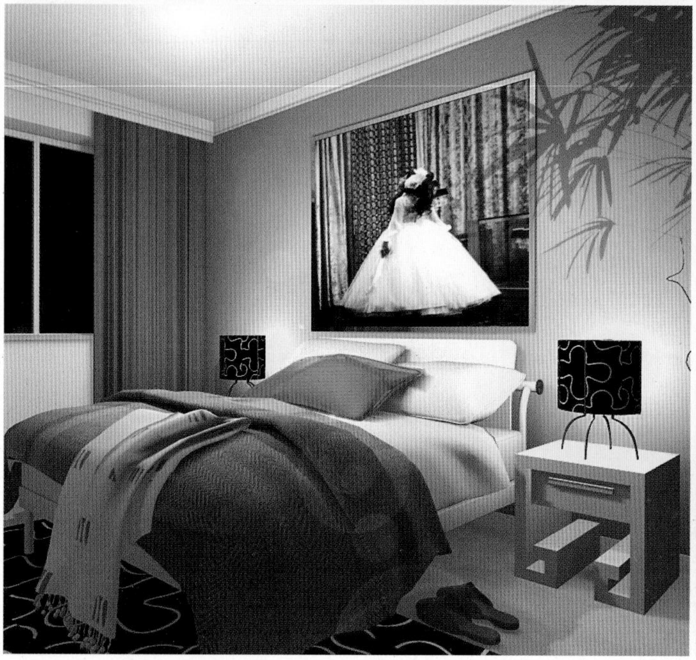

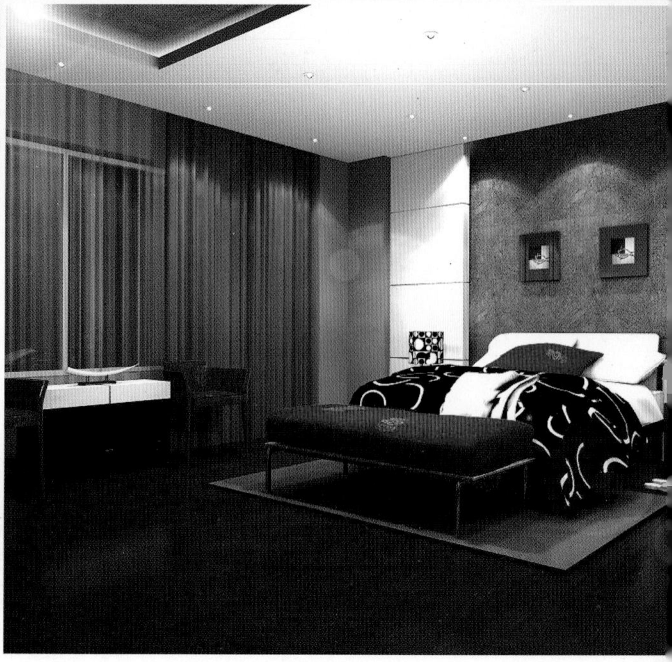